二十四节气与生活

TWENTY-FOUR SOLAR TERMS IN DAILY LIFE

（中英文版）

袁炳富◎编著

高婷婷 葛灵知◎译

图书在版编目(CIP)数据

二十四节气与生活:汉英对照/袁炳富编著. —合肥:安徽大学出版社,2019.5
ISBN 978-7-5664-1803-6

Ⅰ. ①二… Ⅱ. ①袁… Ⅲ. ①二十四节气—普及读物—汉、英 Ⅳ. ①P462—49

中国版本图书馆 CIP 数据核字(2019)第 051585 号

二十四节气与生活(中英文版)　　袁炳富　编著

出版发行:北京师范大学出版集团
安徽大学出版社
(安徽省合肥市肥西路 3 号 邮编 230039)
www.bnupg.com.cn
www.ahupress.com.cn
印　　刷:安徽昶颉包装印务有限责任公司
经　　销:全国新华书店
开　　本:148mm×210mm
印　　张:5.625
字　　数:130 千字
版　　次:2019 年 5 月第 1 版
印　　次:2019 年 5 月第 1 次印刷
定　　价:24.00 元
ISBN 978-7-5664-1803-6

策划编辑:李　梅　韦　玮　　装帧设计:丁　健
责任编辑:韦　玮　李　雪　　美术编辑:李　军
责任印制:赵明炎

前言

我国的“二十四节气”,已于2016年11月30日被正式列入联合国教科文组织人类非物质文化遗产代表作名录。

二十四节气是指24个时节和气候,是我国古代订立的一种用来指导农事的补充历法,是劳动人民长期经验的积累和智慧的结晶。由于古代我国是一个农业社会,农民需要充分了解太阳运行情况,农事完全根据太阳运行方位进行安排,所以在历法中加入了反映太阳运行周期的“二十四节气”,用作确定闰月的标准。二十四节气指导着传统农业生产和日常生活,是中国传统历法体系及其相关实践活动的重要组成部分。

二十四节气的命名反映了季节、气候现象、气候变化等内容。因此,二十四节气又可以划分为四类:

一是表示寒来暑往变化:立春、春分、立夏、夏至、立秋、秋分、立冬和冬至8个节气;

二是象征温度变化:小暑、大暑、处暑、小寒和大寒5个节气;

三是反映降水量:雨水、谷雨、白露、寒露、霜降、小雪和大雪7个节气;

四是反映物候现象或农事活动:惊蛰、清明、小满和芒种4个节气。

白露、寒露和霜降3个节气表面上反映的是水汽凝结和凝华现象，但实质上反映了气温逐渐下降的过程和程度。

惊蛰和清明反映的是自然物候现象，尤其是惊蛰，它用天上初雷和地下蛰虫的复苏来预示春天的回归。

“立”表示一年四季中每一个季节的开始，立春、立夏、立秋和立冬分别对应春、夏、秋和冬四季。由于我国地域辽阔，具有非常明显的季风性和大陆性气候，各地天气气候差异巨大，因此，不同地区的四季变化也有很大差异。立春、立夏、立秋和立冬亦合称为“四立”。公历上一般在每年的2月4日、5月5日、8月7日和11月7日前后。“四立”表示的是天文季节的开始，从气候上说，一般还在上一季节，如立春黄河流域仍处于隆冬之中。

春分、秋分、夏至和冬至是从天文角度来划分的，反映了太阳高度变化的转折点。

“分”在这里表示平分的意思。春分和秋分合称为“二分”，表示昼夜长短相等。这两个节气一般在每年公历的3月20日和9月23日左右。春分日和秋分日，黄道和赤道平面相交，此时黄经分别为0°、180°，太阳直射赤道，昼夜相等。

“至”是极和最的意思。夏至和冬至合称为“二至”，表示夏天和冬天的极致。夏至日和冬至日一般在每年公历的6月21日和12月22日。夏至，太阳直射北纬23°26′，黄经90°，北半球白昼最长。冬至，太阳直射南纬23°26′，黄经270°，北半球白昼最短。

二十四节气的含义为：立春，“立”是开始的意思，立春就是春季的开始；雨水，降雨开始，雨量渐增；惊蛰，“蛰”是藏的意思，惊蛰是指春雷乍动，惊醒了蛰伏在土中冬眠的动物；春分，“分”是平分的意思，

春分表示昼夜平分；清明，天气晴朗，草木繁茂；谷雨，雨生百谷，雨量充足而及时，谷类作物茁壮成长；立夏，夏季的开始；小满，麦类等夏熟作物籽粒开始饱满；芒种，麦类等有芒作物成熟；夏至，炎热的夏天来临；小暑，“暑”是炎热的意思，小暑就是气候开始炎热；大暑，一年中最热的时候；立秋，秋季的开始；处暑，“处”是终止和躲藏的意思，处暑是表示炎热的暑天结束；白露，天气转凉，露凝而白；秋分，昼夜平分；寒露，露水而寒，将要结冰；霜降，天气渐冷，开始有霜；立冬，冬季的开始；小雪，开始下雪；大雪，降雪量增多，地面可能积雪；冬至，寒冷的冬天来临；小寒，气候开始寒冷；大寒，一年中最冷的时候。

本书从二十四节气的由来和相关的物候气候、农事活动、民间风俗等方面，对二十四节气进行系统的介绍和解读，帮助读者了解我国独特的历法分类，以及节气对生活、农事生产等方面的影响。

Preface

The Twenty-four Solar Terms, which originate in China, were officially inscribed on the UNESCO's Representative List of the Intangible Cultural Heritage of Humanity on November 30th, 2016.

The Twenty-four Solar Terms, as a complementary calendar system invented and enacted in ancient China, refer to the twenty-four periods and climates which are used to direct agricultural arrangements and farming activities. It is the crystallization of the accumulated achievements and wisdom of the long-term experience of the Chinese working people. In ancient China, then an agricultural society, farming affairs were carried out entirely according to the sun. The operation of sun became such a key factor that the ancient Chinese divided the sun's annual circular motion into 24 segments and added them to the calendar for determining leap month. The Twenty-four Solar Terms, guiding the traditional agricultural production and people's daily life, play an important part in the traditional Chinese calendar system and its relative practices.

The naming of the Twenty-four Solar Terms reflects the variation of natural phenomenon such as seasons, climate, climate change and so on. Thus, the Twenty-four Solar Terms can be classified into the following four categories:

One group shows the seasonal changes (including eight terms): Start of Spring, Spring Equinox, Start of Summer, Summer Solstice, Start of Autumn, Autumn Equinox, Start of Winter and Winter Solstice;

One represents temperature variations(including five terms): Slight Heat, Great Heat, End of Heat, Slight Cold and Great Cold;

The seven solar terms of Rain water, Grain Rain, White Dew, Cold Dew, Frost's Descent, Slight Snow and Great Snow reflect precipitation phenomenon, indicating the time and intensity of rainfall and snowfall;

The rest embodies natural phenomena or agricultural activities (including four terms): Awakening of Insects, Pure Brightness, Grain Buds and Grain in Ear.

White Dew, Cold Dew and Frost's Descent reflect the condensation and dehydration of water vapor on the surface, but actually show the process and degree of gradual decline in temperature.

Awakening of Insects and Pure Brightness reflect natural phenomena. The former especially utilizes the early thunder and the awakening of hibernated insects to indicate the return of spring.

Li in Chinese means the beginning of each season throughout the year. *Lichun* (Start of Spring), *Lixia* (Start of Summer), *Liqiu* (Start of Autumn) and *Lidong* (Start of Winter) refer to the beginning of the four seasons. China is characterized by a continental monsoon climate due to its vast territory. The climate varies from place to place, so the four seasons in different regions also vary greatly. *Lichun* as well as *Lixia*, *Liqiu* and *Lidong* make up "Four *Li*", falling on around February 4th, May 5th, August 7th and November 7th respectively in the solar calendar. "Four *Li*" represent the beginning of astronomical seasons. In terms of climate, it is still in the previous season generally. For example, the Yellow River basin area is still in the depths of winter when *Lichun* arrives.

Chunfen (Spring Equinox), *Qiufen* (Autumn Equinox), *Xiazhi* (Summer Solstice) and *Dongzhi* (Winter Solstice) are divided from the perspective of astronomy, reflecting the turning point of variation in height of the sun.

Fen here means equal division. *Chunfen* and *Qiufen* constitute "Two *Fen*", signifying the lengths of day and night are equal. These two solar terms usually begin around March 20th and September 23rd annually in the solar calendar. On *Chunfen* and *Qiufen*, the ecliptic and the equatorial plane intersect. At this time, the sun reaches ecliptic longitude of 0° and 180° respectively. Since the sun directly shines on the equator, the periods of day and night are equal in length.

Zhi means the extreme. *Xiazhi* (Summer Solstice) and *Dongzhi* (Winter Solstice) form "Two *Zhi*", representing the extreme of summer and winter. These two solar terms generally fall on June 21st and December 22nd according to the solar calendar. On the Summer Solstice, the sun appears directly overhead at 23°26′ north latitude and moves to the celestial longitude of 90°. The Northern Hemisphere has the longest daytime of the year. On the Winter Solstice, the sun shines directly overhead at 23°26′

south latitude and reaches celestial longitude of 270°. The Northern Hemisphere has the shortest daytime of the year.

To sum up, the meaning of the Twenty-four Solar Terms are as follows: *Li* refers to the beginning, so *Lichun* (Start of Spring) means the beginning of spring; *Yushui* (Rain Water) marks the beginning of precipitation and the increase in rainfall from then on; *Jingzhe* (Awakening of Insects), *Zhe* signifies hiding and during *Jingzhe*, the spring thunder will wake up the hibernated insects; *Chunfen* (Spring Equinox), *fen* means equal division and in *Chunfen* the periods of day and night are equal in length; *Qingming* (Pure Brightness) is characterized by warm weather and lush vegetation; *Guyu* (Grain Rain) means the rain moistens crops and the grain crops can thrive with enough and timely rainfall; *Lixia* (Start of Summer) means the beginning of summer; *Xiaoman* (Grain Buds) signifies summer crops grow well, but not mature; *Mangzhong* (Grain in Ear) means wheat and other crops become mature; *Xiazhi* (Summer Solstice) means the hot summer arrives; *Xiaoshu* (Slight Heat), *Shu* means hotness in Chinese, so the weather become hot in *Xiaoshu*; *Dashu* (Great Heat) signifies the hottest time of the year; *Liqiu* (Start of Autumn) signifies the beginning of autumn; *Chushu* (End of Heat), *Chu* refers to termination and hiding in Chinese, so *Chushu* marks the end of hot summer days; *Bailu* (White Dew) means the weather gets cool and dewdrops glisten on the grass; *Qiufen* (Autumn Equinox) means equal lengths of day and night in autumn; *Hanlu* (Cold Dew) signifies the weather becomes cold enough to reach dew point; but not

cold enough to reach frost point; *Shuangjiang* (Frost's Descent) signifies the weather becomes cold and frost begins to form; *Lidong* (Start of Winter) means the start of winter; *Xiaoxue* (Slight Snow) marks the beginning of snowfall; *Daxue* (Great Snow)means more and heavy snow and the ground is covered with snow; *Dongzhi* (Winter Solstice) means the cold winter comes; *Xiaohan* (Slight Cold) signifies the weather becomes cold; *Dahan* (Great Cold) brings the coldest period of the year.

This book gives a systematic introduction and explanation of the Twenty-four Solar Terms in terms of its origin and relative knowledge on weather and climate, phonological phenomena, agricultural activities and folk customs, in order to help the readers to gain further understanding on this unique calendric system and its great impact on several aspects of people's life and agricultural production.

目录

Contents

第一部分
二十四节气的由来

part one

The Origin of the
Twenty-four Solar Terms

早在春秋战国时期，我国古代利用土圭实测日晷（即在地面上竖一根杆子来测量正午太阳影子的长短），以确定春分、夏至、秋分、冬至4个节气。一年中，土圭在正午时分影子最短的一天为夏至（又称“日短至”“短至”），最长的一天为冬至（又称“日长至”“长至”），影子长度适中的为春分或秋分。春秋时期的著作《尚书》中就对节气有所记述。在商朝时只有4个节气，到了周朝时发展到了8个。“二十四节气”的名称首见于西汉刘安的《淮南子·天文训》，《史记·太史公自序》的《论六家要旨》中也曾提到阴阳、四时、八位、十二度、二十四节气等概念。我国古代用农历（月亮历）计时，用阳历（太阳历）划分春夏秋冬和二十四节气。我们祖先把五天叫“一候”，三候为一气，称“节气”，全年分为七十二候和二十四节气。

太阳从黄经0°起，沿黄经每运行15°所经历的时日称为“一个节气”。每年运行360°，共经历24个节气，每月2个。其中，每月第一个节气为“节气”，即：立春、惊蛰、清明、立夏、芒种、小暑、立秋、白露、寒露、立冬、大雪和小寒等12个节气；每月的第二个节气为“中气”，即：雨水、春分、谷雨、小满、夏至、大暑、处暑、秋分、霜降、小雪、冬至和大寒等12个节气。“节气”和“中气”交替出现，各历时15天，后来人们逐渐把“节气”和“中气”统称为“节气”。

公元前104年，由邓平、唐都、落下闳等人制定的《太初历》正式把二十四节气定于历法，明确了二十四节气的天文位置。二十四节气依次为：立春、雨水、惊蛰、春分、清明、谷雨、立夏、小满、芒种、夏至、小暑、大暑、立秋、处暑、白露、秋分、寒露、霜降、立冬、小雪、大雪、冬至、小寒、大寒。

这就是二十四节气的形成过程。

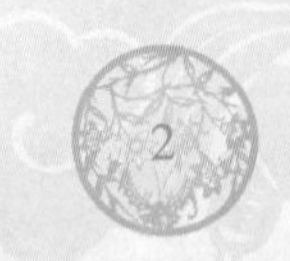

As early as the Spring and Autumn Period and Warring States Period, ancient Chinese people first determined the four seasons as spring, summer, autumn and winter by measuring the length of the sun's shadow (a stick vertically placed on the ground to measure the length of sun's shadow at noon) on an ancient timekeeper instrument named *tugui*. At that time, people defined the days with the shortest and longest sun's shadow on *tugui* as the Summer Solstice (also known as *Duanzhi* or *Ri Duanzhi*) and Winter Solstice (also called *Changzhi* or *Ri Changzhi*). When the sun's shadow was medium, it might fall on the Spring Equinox or Autumn Equinox. The solar terms were mentioned in the *Book of Documents* of the Spring and Autumn Period. In the Shang Dynasty, there were only four solar terms and by the Zhou Dynasty the number increased to eight. The names of Twenty-four Solar Terms were first recorded in Liu An's *Huainanzi: Patterns of Heaven* during the early Western Han Dynasty. *On Principles of Six Mentors* of *Taishi Gong's Preface* of *the Grand Historian of China* also gave an account of yin and yang, the Four Seasons, the Eight Diagrams, the Duodenary Series and the Twenty-four Solar Terms. In ancient China, the lunar calendar was used to record time while the solar calendar became criterion for dividing the four seasons and the Twenty-four Solar Terms. Our ancestors referred to five days as one *hou* and addressed three *hou* as one *qi*, also known as *jieqi*. The whole year was divided into seventy-two *hou* and twenty-four solar terms.

From the starting point 0° ecliptic longitude, every 15° the sun's movement is regarded as one solar term. The 360° revolution of the earth is equally divided into 24 segments which are known as the Twenty-four Solar Terms, with two solar terms per month. Among them, the first solar terms of each month are called *jieqi*, namely, Start of Spring, Awakening of Insects, Pure Brightness, Start of Summer, Grain in Ear, Slight Heat, Start of Autumn, White Dew, Cold Dew, Start of Winter, Great Snow and Slight Cold; the second solar terms of each month are addressed as *zhongqi*, including Rain Water, Spring Equinox, Grain Rain, Grain Buds, Summer Solstice, Great Heat, End of Heat, Autumn Equinox, Frost's Descent, Slight Snow, Winter Solstice and Great Cold. *Jieqi* and *zhongqi* appear alternately, each of which lasts 15 days. So far *jieqi* and *zhongqi* have been collectively called *jieqi* by ancient Chinese.

In 104 BC, Taichu Calendar formulated by Deng Ping, Tang Du, Luo Xiahong and other people formally added the Twenty-four Solar Terms to the calendar and determined the astronomical position of the Twenty-four Solar Terms. The Twenty-four Solar Terms are Start of Spring, Rain Water, Awakening of Insects, Spring Equinox, Pure Brightness, Grain Rain, Start of Summer, Grain Buds, Grain in Ear, Summer Solstice, Slight Heat, Great Heat, Start of Autumn, End of Heat, White Dew, Autumn Equinox, Cold Dew, Frost's Descent, Start of Winter, Slight Snow, Great Snow, Winter Solstice, Slight Cold and Great Cold.

This is the formation of the Twenty-four Solar Terms.

第二部分

历法的基本概念

part two

The Basic Concept of Calendric System

据《现代汉语词典》，历法是指用年、月、日来计算时间的方法。主要分阳历、阴历和阴阳历3种。

阳历也叫“太阳历”，它是以地球绕太阳一周的时间（365.24219天）为一年，平年365天，闰年366天，一年分12个月。

阴历也叫“太阴历”，它是月球绕地球一周的时间（29.53059天）为1个月，大月30天，小月29天，积12个月为一年，一年354天或355天。伊斯兰教历就是阴历的一种。

阴阳历是以月球绕地球一周的时间为1个月，但设置闰月，使一年的平均天数跟太阳年的天数相符，因此这种历法与月相相符合，也与地球绕太阳的周年运动相符合。农历就是阴阳历的一种。具体的历法还包括纪年的方法，参看公历、农历、伊斯兰教历。

According to the *Modern Chinese Dictionary*, a calendric system refers to the method of calculating time by days, months and years. It is mainly divided into three types: the solar calendar (*yangli*), the lunar calendar (*yinli*) as well as the lunisolar calendar (*yinyangli*).

The solar calendar, based on the sun's movement, regards the time the earth rotates around the sun once as a solar year (365.24219 days). A year can be divided into 12 months, with 365 days for a common year and 366 days for a leap year.

The lunar calendar is a calendar based upon the monthly cycles of the moon's phases (29.53059 days). Since each lunation is approximately 29.5

days, it is common for the months of a lunar calendar to alternate between 29 and 30 days. Since the period of twelve such lunations, a lunar year, is only 354 or 355 days. The Islamic calendar is a kind of the lunar calendar.

The lunisolar calendar sees the time the moon rotates around the earth once as a month. At the same time, a leap month is also listed in the lunisolar calendar, making the medium days of a year in conformity with the days of solar year. Therefore, this calendar is consistent with the moon phase and the annual movement of the earth around the sun. In fact, the Chinese lunar calendar is a kind of lunisolar. The specific calendar also includes the method of annals, referring to the solar calendar, the lunar calendar and the Islamic calendar.

历的起源

"历"的设立,是为了判别节候、记载时日、规定计算时间的标准。太古时,人们"日出而作,日入而息",并不需要历法。当人类逐渐进化,关系日益复杂,应有一种标准来计量时间单位,这种单位必须采取所经时间的固有事象为依据,而且需要人们达成共识。于是以一昼夜为一日,为一般历法的基本单位;以月球盈昃为一月,为太阴历的基础;以一寒暑为一年,为太阳历的基础。

相传,"天皇氏制干支,伏羲氏作甲历,黄帝氏命大挠作甲子,太昊氏设历正、颛顼氏作新历,帝尧氏命羲和敬授人时,期三百有六旬

有六日，以闰月定四时成岁，与近世推算回归年略相等。夏后氏颁夏时，为我国正朔之唯一标准。”

夏商周的历制不同：夏以建寅之月为岁首，商以建丑之月为岁首，周以建子之月为岁首。秦以建亥之月为岁首，汉初保留秦制未改，武帝时改正朔为夏正。王莽改用殷正、建丑，其后魏明帝、唐武后及肃宗，先后改朔，但不久后仍用夏正，一直到清末。

清咸丰四年（1854年）太平天国修改历法，以366日为一年，一年12个月，单月30日，双月31日，以干支纪日，与中历相同，礼拜顺序也与西方历法一致，但是将节置于月首，气置于月中，过了14年后被废除。

汉太初一直到清末的2 000多年间，大部分时间都以建寅为岁首，中间虽改为正朔，但最多10余年，最少1、2年之后，仍采用夏正。

古代希腊的历法，与我国旧历最相近，也是采用太阴历。罗马人建国时所定下的历法，一年为10个月，共304日。公元前46年，罗马大帝恺撒命令执政官改正历法，现今所用的太阳历即是在此历法的基础上进行修订的。公元1582年经罗马法王格列高利改正，即为现在世界各国通行的历法。

The Origin of Calendar

The calendar is used to determine the solar terms, record time and specify the criterion for calculating the time. People in the remote antiquity worked at sunrise and rested at sunset so they didn't need a calendar. When human society gradually developed and the relationship had become more and more complex, there should be a unit of time measurement as the criterion. This unit must be based on inherent phenomenon of certain things during some time and recognized by people. Hence, the basic unit of general calendar stipulates that a nychthemeron can be regarded as one day; the basic unit of lunisolar calendar rules that one cycle of the moon movement can be deemed as one month; one round of summer and winter being one year forms the base of lunisolar calendars.

It is said that Heavenly Sovereign (Tianhuang) devised sexagenary cycle and Fuxi created *jiali*. Yellow Emperor (Huangdi) ordered Danao to formulate *jiazi* and his grandson Zhuanxu established a new calendar. Taihao set up the post in charge of astronomical calendar. Emperor Yao commanded Xihe to notify common people of calendar, making them do the farm work in the right season. There are 366 days in the fixed period and the four seasons are determined by leap month to constitute *sui*, roughly equals to tropical year of the modern times. Xiahou issued *xiashi*, becoming the only standard of *zhengshuo* (the first day of the first month in Chinese) in China.

The calendric systems of these three dynasties were different: the month of *jianyin* was deemed as the beginning of the year in Zhou Dynasty; the month of *jianchou* the beginning of the year in Shang Dynasty; the month of *jianzi* the beginning of the year in Zhou Dynasty. In Qin Dynasty, the beginning of the year was the month of *jianhai* and this practice continued until the early Han Dynasty. When Emperor Wu of Han ascended the throne, he began to abolish it and adopt *xiazheng*.

Wang Mang at the end of Han Dynasty substituted *yinzheng* and *jianchou*. Afterwards, Emperor Ming of Northern Wei, Wu Zetian and Emperor Su of Tang Dynasty changed the first day of each lunar month. But soon *xiazheng* was reused for a long time until the end of Qing Dynasty.

The leader of the Taiping Heavenly Kingdom also mandated a reform to calendar in 1854, taking 366 days for a year and a year for 12 months. There were 30 days in odd month and 31 days in even month. In addition, the sexagenary cycle was the same as *zhongli*, and its week order was also identical to western custom. *Jie* (sectional terms) were placed at the beginning of the month and *qi* (middle terms) in the middle. This calendar lasted 14 years and was finally abandoned.

The month of *jianyin* had been the beginning of the year over 2 000 years from Taichu period of Han Dynasty to the end of Qing Dynasty. Although some emperors during the above time employed *zhengshuo* (the first day of the first month), which lasted ranging from 1 or 2 years to over 10 years, *xiazheng* was still in use not long after.

The ancient Greek calendar is close to the old Chinese calendar, both adopting lunisolar calendars. When Romans established their country, they formulated the calendar. According to their calendar, there were ten months in one year, reaching up to 304 days. In 46 BC, Roman Emperor Julius Caesar ordered his praetor to reform this calendar, which is the base of lunisolar calendar used nowadays. Pope Gregory XIII improved and enacted the Gregorian calendar in 1582, which turned to be the current common calendar used by all countries in the world.

阳 历

年有3种:恒星年、近点年和回归年。

地球绕日一周,历时365天6小时9分9秒,称为“恒星年”;太阳过近地点循黄道东行一周,复过近地点,历时365天6小时13分48秒,称为“近点年”;太阳过春分,循黄道东行一周,复过春分点,历时365天5小时48分46秒,称为“回归年”,也称“岁实”。

因二分点(春分点和秋分点)每年沿黄道向西逆行约50秒,所以回归年较恒星年的时间更短,相差20分23秒,被称为“岁差”。

此三种年的时间不同,若想每年之节气寒暑不变,则要取回归年为制历之年。对于回归年的时间,有一歌诀如下,可方便记忆:

地球绕日一周年,要知时间有多少?三六五日加五小时四十八分四六秒。

自1月1日至次年1月1日叫一年,年长本应与“岁实”相等,但是一年的日数,必须是整数,不便将奇零之时数计入,故以365日为一年,每年余5小时48分46秒,积至四年约满一日,故每4年增加一日,为闰日,谓之“闰年”。其无闰日之年,谓之“平年”,平年365日,闰年366日。

置闰之法,为便利起见,按公元计算:凡是公元年数能以4除尽的年份皆为闰年;若以百除之得整数,再以4除之而除不尽的年份,都不置闰,能除尽者则仍为闰年。

地球的轨道为椭圆形，故距日有远近。1月1日，其距离最近，谓之“近日点”；7月2日距离最远，谓之“远日点”。一年的开始，谓之“岁首”，亦称“年始”，阳历以近日点为岁首，为元月一日。

阳历有月大月小之分。阳历每年分12个月，每月的日数不规则，月大31天，月小30天，平年2月28天，闰年2月29天。阳历的1个月，与月球运行无关。除2月有平年闰年之分外，每年各月的天数都是确定的：7月以前，单月是31天，双月是30天；8月以后，双月是31天，单月30天。

The Solar Calendar

There are three kinds of years: sidereal year, anomalistic year and tropical year.

The sidereal year is the time taken by the earth to orbit the sun once with respect to the fixed stars. Its average duration is 365 days, 6 hours, 9 minutes and 9 seconds. The anomalistic year is the time taken for the earth to complete one revolution with respect to its apsides. Its average duration is 365 days, 6 hours, 13 minutes and 48 seconds. The tropical year, also known as *suishi*, is defined as the period of time for the mean ecliptic longitude of the sun to increase by 360°. The mean tropical year is approximately 365 days, 5 hours, 48 minutes and 46 seconds.

The precession of the equinoxes (also called *suicha*) refers to the slightly earlier occurrence of the equinoxes each year due to the slow continuous westward shift of the equinoctial points along the elliptic by 50

seconds of arc per year. Therefore, the tropical year is about 20 minutes and 23 seconds shorter than the sidereal year.

The time of the above-mentioned three kinds of years is different. In order to make the annual solar terms and seasons unchanged, the calendar of the tropical year is chosen as the standard. There is a need for us to memorize the time of the tropical year. For the sake of easy memory, versified formula is composed as follows.

How much time does it take when the earth completes one revolution around the sun? The answer is 365 days, 5 hours, 48 minutes and 46 seconds.

The time from January 1st of this year to January 1st of the following year is termed one year and the length of one year is equal to *suishi*. The total days of a year must be integer. It is inconvenient to include extra part. Hence, 365 days can be seen as a year. The remaining 5 hours, 48 minutes and 46 seconds of each year to be added in four years reach approximately one day. Therefore, one day is accumulated every four years and addressed as a leap day. The year with the leap day is leap year, which has 366 days, as opposed to common year without the leap day, which has 365 days.

For the sake of convenience, the identification of leap year can be calculated according to AD. The year evenly divided by 4 belongs to leap year; if the year can be evenly divided by 100, it is not a leap year; if the year can be evenly divided by 400, it is a leap year.

The orbit of the earth is elliptical, so the point in the orbit can be far

or near to the sun. In January 1st, this point, usually termed perihelion, is nearest to the sun. It is the opposite of aphelion, which is the farthest point from the sun in July 2nd. The beginning of the year is called *suishou* in Chinese, also known as *nianshi*. In the solar calendar, *suishou* falls on January 1st. That is the time perihelion appears. In the solar calendar, the time perihelion appears is regarded as *suishou*, falling on January 1st.

The solar calendar has "big" months and "small" months. Each year can be divided into 12 months in the solar calendar and the number of days per month is irregular. There are 31 days in the "big" months and 30 days in the "small" months. February has 28 days in common year and 29 days in leap year. One month of the solar calendar has nothing to do with the operation of the moon. In addition to the distinction between common year and leap year in February, the number of days each month is certain: before July, the odd month has 31 days and the even month has 30 days; after August, there are 31 days in even month and 30 days in odd month.

阴历

月球运行的轨道，称为“白道”，白道与黄道同为天体上的两大圆，以 5°9′分而斜交，月球绕地球一周，出没于黄道两次，历时 27 日 7 小时 43 分 11 秒半，为月球公转一周年所需的时间，谓之“恒星月”。只有当月球绕地球的时候，地球因公转而位置有所变动，计前进 27°，

而月球每日行13°15′,所以月球自合朔,绕地球一周,再至合朔,实需约29天12小时44分3秒,谓之"朔望月",习俗所谓"一月",即指朔望月。因每月天数不能有奇零,故阴历一个月为29天或30天。每月以合朔之日为首,即以朔日为初一日。每年以接近立春之朔日为岁首。

地球绕日一周的时间,即月球绕地球12⅓周的时间。一年内月数不能有奇零,故一年12个月,仅354日,与岁实相较,约余11日,积至3年,余33日,故每3年须置一闰月,尚余3日或4日,再积2年,共余25日或26日,可置一闰月。平均计算,每19年须置七闰。以有节无气之月为闰月,有闰月之年为闰年,闰年有13个月;无闰月之年为"平年",平年则12个月。

The Lunar Calendar

The moon's orbit and ecliptic are the two circles of the celestial body. Its orbital plane is inclined by about 5°9′ with respect to the ecliptic plane. The moon completes one revolution around the earth in 27 days, 7 hours, 43 minutes and 11 seconds, passing the ecliptic twice. This time can also be called sidereal month. While the moon is orbiting the earth, the earth is progressing in its orbit around the sun and moving to a new position by over 27° compared to the past. In addition, the moon travels by 13°15′ every day. Therefore, The moon must move for about 29 days, 12 hours, 44 minutes, 3 seconds to realign with the sun, which is addressed as synodic month. One month in traditional sense is in conformity with synodic

month. Since the number of days each month must be integer, each month in lunar calendar has 29 or 30 days. The day of the Sun-Moon-Earth aligned marks the start of each month, that is, *shuori* becomes the first day of each month. What's more, the *shuori* close to the Start of Spring in time is defined as the beginning of the year.

The period the earth completes one revolution relative to the sun is the same as the time the moon orbits around the earth twelve and one third times. The number of months each year must be integer, so there is 12 months, reaching up to 354 days. It has extra 11 days compared to *suishi*. The remaining 11 days to be added in three years can reach 33 days. Therefore, one leap month appears every three years, with a remainder of 3 or 4 days. In the following two years the remainder can rise to 25 or 26 days and another leap month turns up. According to average calculation, there develops 7 leap months every 19 years. The month containing *jieqi* but without *zhongqi* belongs to leap month and the year with leap month is leap year. The leap year has 13 months while the common only includes 12 months.

阳阴历之异同

“阳历”又名“太阳历”，是以地球绕行太阳一周为一年，为西方各国所通用，所以又叫“西历”。我国自民国元年采用阳历，故又名“国

历”。为与我国旧有之历相区别,故又名“新历”。

“阴历”又名“太阴历”,是以月球绕行地球一周为一月,再配合地球绕日一周之时数为一年,实际上等于阴阳合历。我国在民国纪元前采用此历,为与现行历法相区别,故名“旧历”。一般,人们认为阴历适于农家,而名之曰“农历”,实际上并不如此。

地球绕日一周,历 365 日 6 时 9 分 9 秒。自春分回至春分,需要 365 日 5 时 48 分 46 秒,称为“岁实”。因春分点逐渐西行,所以岁实较地球周天的时刻更短,相差约 20 分 23 秒,称为岁差,自正月一日至次年正月一日,谓之“年”。然而一年之内,不能有奇零时数,故以 365 日为平年。每年所余 5 时 48 分 46 秒,积至 4 年约满 1 日,故每过 3 年,增加 1 日,为闰年。

以前历代的历法,虽说法不一,但是其要旨基本相同。无非就是,阳历的月份,仅为年的分段,与晦朔弦望无关,所以日数可以规定。阴历的月份,则是以日月合朔之日为首,两次合朔相距 29 日有半,所以一个月的日数,或为 29,或为 30。因月份规定不同,所以年法也不一样。阴历年以近立春之朔日为始,一年之内,月数不能有奇零,然而 12 个月后,仅得 354 日,以之为年,与岁实相差约 11 日。积至 3 年,已少 33 日,所以须每 3 年置一闰月。再积 2 年,又少 25 日,也可置一闰。平均计算,每 19 年,须置七闰。

The Similarities and Differences Between the Solar Calendar and the Lunar Calendar

The solar calendar, also known as *taiyangli* in Chinese, defines the time the earth travels around the sun once as one year. It is commonly used by western countries, also called *xili* in Chinese. China has adopted the solar calendar since the first year of the Republic of China. Thus, it gains a new name *guoli* (national calendar) in Chinese. In order to be symmetrical with Chinese old calendar, it is also named *xinli* (new calendar).

The lunar calendar, also known as *taiyinli*, defines the time the moon completes one revolution around the earth as one month. Coupled with the time the earth moves around the sun once, it actually amounts to the lunisolar calendar. Our country adopted this calendar before the founding of the Republic of China. In order to be distinguished from the current calendar, it is named *jiuli* (old calendar) in Chinese. Most people think that the lunar calendar is suitable for farmers, so it is called *nongli* (farmers' calendar) in Chinese. In fact, that is not the case.

The earth makes one revolution around the sun in 365 days, 6 hours, 9 minutes and 9 seconds. The time interval between two sequential passes of the sun's true center through the point of the Spring Equinox is called *suishi*, reaching up to 365 days, 5 hours, 48 minutes and 46 seconds. Because of the slow continuous westward shift of the spring equinoctial

point, the tropical year is 20 minutes and 23 seconds less than the period for the earth to return to the same point, which is defined as the precession of the equinoxes. One year is the time ranging from the first day of the first month this year to the first day of the first month the following year. The number of days each year must be integer, so 365 days constitute common year. The remaining 5 hours, 48 minutes and 46 seconds of each year to be added in four years reach approximately one day. Therefore, one day is accumulated every other three years and addressed as the leap day. The year with the leap day is a leap year.

Although the calendars of previous dynasties are different in stipulation, they have a unified gist. The difference lies in the counting of the month. The month of the solar calendar is merely the sections of year. It has nothing to do with the revolution and rotation of the moon, so the number of days can be fixed. The first day of the month in the lunar calendar is characterized by the day of *heshuo* (the moon and the sun are at the same degree on the celestial sphere). The two successive days of *heshuo* are 29 and half days apart. Therefore, the number of days each month can be 29 or 30 days. Because of the different month determinations, there are diverse year determinations. The *shuori* that is close to the Start of Spring marks the start of the year. The number of months must be integer and there are 12 months in a year, reaching up to 354 days. It is shorter than *suishi* by 11 days. It would be 33 days apart in 3 years. Therefore, one leap month appears every three years. In the following two years the difference in time can be 25 days and another leap month turns up. According to average calculation, 7 leap months are developed every 19 years.

第三部分

二十四节气与十二星座

part three

The Twenty-four Solar Terms and the Twelve Constellations

前面已讲了，一年有 12 个月、24 个节气，每一个月都有 2 个节气，其中一个是节，一个是气。

那么，什么是“节”呢？节等于地球在虚空的转动，也叫“节气”。那什么又是“气”呢？气是地球本身内在的放射功能，太阳、月亮跟自然界的放射是配合的，也叫“中气”。古时把节气称“气”，每月有两个气，前一个气叫“节气”，后一个气叫“中气”。农历闰年的闰月仅有一个节气，没有“中气”。一般而言，“节气”与“中气”交替出现，间隔 15 天。在农历中，立春为二十四节气的第一个节气。

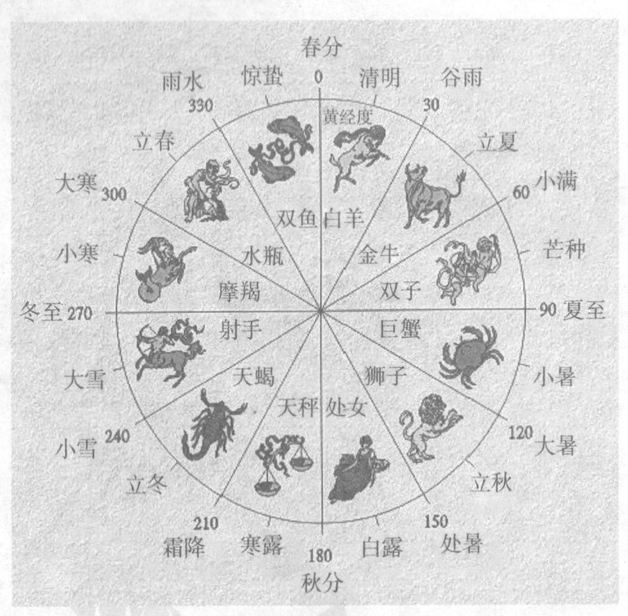

二十四节气与十二星座对应图

此处所讲的十二星座指的就是气，也是后一个气——“中气”。十二星座以春分为起始，太阳每年到春分点那个时刻开始计算星座。

每年春分时候开始的星座就是十二星座第一个星座白羊座，谷雨时候进入金牛座，小满进双子座，这三个星座都在春季（以天文学分类法为标准）；夏至就是巨蟹座，大暑即为狮子座，处暑进入处女座，这三个星座经过夏季；秋分来到天秤座，霜降时候进入天蝎座，小雪就是射手座；冬至进入摩羯座，大寒就是水瓶座，雨水进入最后一个星座双鱼座。接着又是春分的白羊座，这样一个循环就是一年。

二十四节气是根据太阳在黄道（即地球绕太阳公转的轨道）上的位置来划分的。太阳从春分点出发，每前进 15° 为 1 个节气；运行一周又回到春分点，为一回归年，合 360°。十二星座是 30° 换一个星座，从春分点的白羊座出发，运行一圈还回到春分点。

十二星座与节气对照情况如下。

白羊座对应春分到谷雨；

金牛座对应谷雨到小满；

双子座对应小满到夏至；

巨蟹座对应夏至到大暑；

狮子座对应大暑到处暑；

处女座对应处暑到秋分；

天秤座对应秋分到霜降；

天蝎座对应霜降到小雪；

射手座对应小雪到冬至；

摩羯座对应冬至到大寒；

水瓶座对应大寒到雨水；

双鱼座对应雨水到春分。

It has been mentioned before that there are twelve months and twenty-four solar terms in a year. Each month has a pair of sectional (*jie*) and middle (*qi*) solar terms interlaced with each other.

What's *jie*? *Jie*, also known as sectional terms, equate to rotation of the earth in virtual space. Then what's *qi*? *Qi*, also addressed as middle terms, refers to the inherent radiation function of the earth itself. The moon and the sun are in harmony with the radiation of nature. In ancient times, there are two solar terms in a month: the one in the beginning of the first half month is called as terms (*jieqi*) and the other one in the middle of the latter half month is known as middle terms (*zhongqi*). With regards to the leap month in lunar leap year, there is only one solar term without the middle term. In general, first terms and middle terms appear respectively with the duration of 15 days. The Twenty-four Solar Terms embody a complete circle of the sun and divide the circle into 24 segments, beginning with the Start of Spring.

The twelve zodiac constellations mentioned here refer to *qi*, namely the latter "middle terms", which were determined in the ancient time by dividing the zodiac (a band around the ecliptic) into 12 equal sectors measuring from the Spring Equinox.

When the Spring Equinox falls, the first constellation turns out to be Aries. Taurus and Gemini fall on the Grain Rain and Grain Buds

respectively. These three constellations are all in spring (defined by astronomy). Cancer appears when the Summer Solstice comes and Leo falls on the Great Heat. After that, the End of Heat brings Virgo. These three constellations are all in summer. For the following two solar terms the Autumn Equinox and the Frost's Descent, Libra and Scorpio arrive. While the Slight Snow approaches, Sagittarius turns up. In winter, the Winter Solstice falls, accompanied by the emergence of Capricorn. The last two constellations Aquarius and Pisces fall on the Great Cold and Rain Water. Then comes the Aries. The cycle of one year finally ends.

The Twenty-four Solar Terms are divided according to the position of the sun on ecliptic (the orbit of the earth around the sun). The sun starts moving from the point of the Spring Equinox, with 15° of the sun's longitude between the terms. It returns to the point of the Spring Equinox after making a complete 360° revolution, which is called the tropical year. These constellations can be divided into 12 equal parts and there appears one constellation every 30°. The first constellation Aries, starting from the point of the Spring Equinox, returns to the same point after finishing one circle.

Comparison table of zodiac constellations and solar terms are listed as following.

Aries corresponds from Spring Equinox to Grain Rain;

Taurus corresponds from Grain Rain to Grain Buds;

Gemini corresponds from Grain Buds to Summer Solstice;

Cancer corresponds from Summer Solstice to Great Heat;

Leo corresponds from Great Heat to End of Heat;

Virgo corresponds from End of Heat to Autumn Equinox;

Libra corresponds from Autumn Equinox to Frost's Descent;

Scorpio corresponds from Frost's Descent to Slight Snow;

Sagittarius corresponds from Slight Snow to Winter Solstice;

Capricorn corresponds from Winter Solstice to Great Cold;

Aquarius corresponds from Great Cold to Rain Water;

Pisces corresponds from Rain Water to Spring Equinox.

第四部分

节气与生活

part four

The Twenty-four Solar Terms in Daily life

一月

一月有小寒、大寒两个节气。

小寒

1. 日期计算

计算公式：(Y×D+C)-L

公式解读：Y= 年数的后 2 位，D=0.2422，L= 闰年数，21 世纪 C=5.4055，20 世纪 C=6.11。

举例说明：1988 年小寒日期 = (88×0.2422+6.11)-[(88-1)÷4] =27-21=6，1 月 6 日小寒。

例外：1982 年计算结果加 1 日，2019 年减 1 日。

2. 节气特点

以农历计算，小寒是二十四节气中的第二十三个节气，时间是在公历 1 月 5 ～ 7 日之间，太阳位于黄经 285°。我国这时正值"三九"前后，小寒标志着一年中最寒冷的日子即将开始。据《月令七十二候集解》："十二月节，月初寒尚小，故云，月半则大矣。"此时天气开始变冷。俗话说："小寒大寒，冷成冰团，幽阖大雪红炉暖。"

我国古代将小寒分为三候："一候雁北乡，二候鹊始巢，三候雉始鸲。"古人认为候鸟中大雁是顺阴阳而迁移，此时阳气已动，所以大雁开始向北迁移；此时北方到处可见到喜鹊，并且感觉到阳气而开

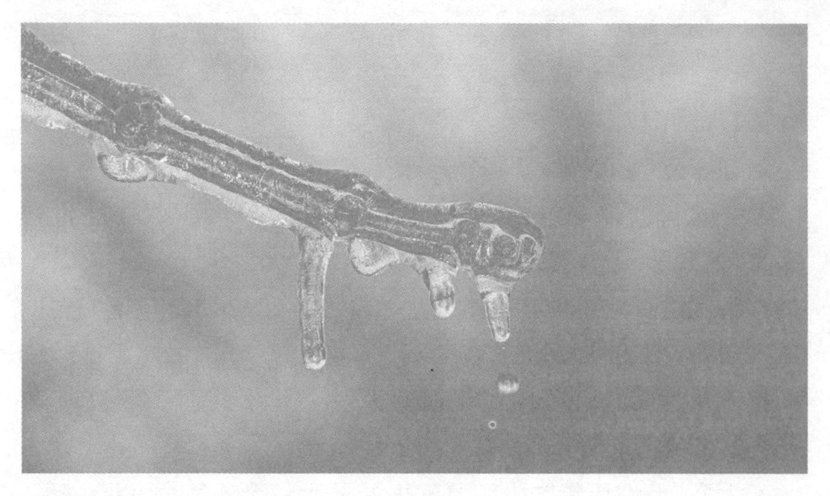

始筑巢；第三候“雉鸲”的“鸲”为鸣叫的意思，雉在接近四九时会感阳气萌动而鸣叫。

3. 节气习俗

【饮食】小寒正是吃麻辣火锅、红焖羊肉的好时节。居民日常饮食偏重于暖性食物，如羊肉、狗肉，其中又以羊肉汤最为常见，有的餐馆还推出当归生姜羊肉汤。近年来，一些传统的冬令羊肉菜肴重现餐桌，再现了寒冬食俗。

【活动】俗话说：“小寒大寒，冷成冰团。”小寒节气正处于“三九”寒天，是人们加强体育锻炼、提高身体素质的大好时机。南方人在小寒有一套具有地域特色的体育锻炼方式，如跳绳、踢毽子、滚铁环等。如果遇到下雪，人们更是欢呼雀跃，打雪仗、堆雪人，这些活动会使全身暖和，血脉通畅。

大寒

1. 日期计算

计算公式：(Y×D+C)-L

公式解读：Y= 年数的后 2 位，D=0.2422，L= 闰年数，21 世纪 C=20.12，22 世纪 C=20.84。

举例说明：2089 年大寒日期 =(89×0.2422+20.12)-[(89-1)÷4]= 41-22=19，1 月 19 日大寒。

例外：2082 年的计算结果加 1 日。

2. 节气特点

大寒，是农历年二十四节气中的最后一个节气。每年公历 1 月 20 日前后，太阳到达黄经 300° 时，即为大寒。《授时通考·天时》引《三礼义宗》："寒气之逆极，故谓大寒。"这时寒潮频繁南下，是我国部分地区一年中的最冷时期，风大，低温，地面积雪不化，呈现

出冰天雪地、天寒地冻的严寒景象。这个时期，铁路、邮电、石油、输电线路、水上运输等部门需要特别注意大风降温、大雪等灾害性天气的预警和防范。

3. 节气习俗

【饮食】大寒时节人们开始忙着除旧布新，腌制年肴，准备年货，因为我国人最重要的节日——春节就要到了。期间还有一个对于北方人非常重要的日子——腊八，即阴历十二月初八。在这一天，人们用米、豆等谷物加上栗子、红枣、莲子等熬成一锅香甜美味的腊八粥。腊八粥起源于佛教，传说释迦牟尼在这一天成道，因此寺院每逢这一天煮粥供佛，此后民间相沿成俗。

华南地区民间还有大寒节瓦锅蒸煮糯米饭的习俗，糯米味甘，性温，食之具有御寒滋补功效。

【活动】在大寒至立春这段时间，有很多重要的民俗和节庆。如尾牙祭、祭灶和除夕等，有时甚至连我国最大的节日春节也处于这一节气中。尾牙源自于拜土地公做“牙”的习俗。

所谓二月二为“头牙”，以后每逢初二和十六都要做“牙”，到了农历十二月十六日正好是尾牙。尾牙同二月二一样有春饼（南方叫“润饼”）吃，这一天买卖人要设宴，白斩鸡为宴席上不可缺的一道菜。

大寒节气里，除干农活外，还要为过年奔波——赶年集、买年货、写春联，准备各种祭祀供品，扫尘洁物，除旧布新。同时祭祀祖先及各路神灵，祈求来年风调雨顺。此外，旧时大寒时节有些地区的街上还常有人争相购买芝麻秸，因为“芝麻开花节节高”，人们将买回的芝麻秸在除夕夜洒在路上，让孩童踩碎，以求来年日子更好。

January

In January, there are two solar terms—Slight Cold and Great Cold.

Slight Cold

1. Calculation Formula of Slight Cold

The calculation formula: (Y×D+C)-L

The interpretation of formula: Y=the latter two digits of the year number, D=0.2422, L= the number of leap year, 21st C=5.4055, 20th C =6.11("C" stands for century.).

For example, the calculating date of Slight Cold in 1988=(88×0.2422+ 6.11)-[(88-1)÷4]=27-21=6.

Therefore, the day would fall on January 6th.

Exceptions: The date of Slight Cold in 1982 would fall on the day after the calculating result, while that for 2019 would fall on the day before the calculating result.

2. Features of Slight Cold

Slight Cold (*Xiaohan*) is the 23rd solar term of the Twenty-four Solar Terms in the traditional Chinese calendar and falls on a day between January 5th and January 7th, when the sun reaches the celestial longitude of 285°. For Chinese people, the Slight Cold is around the third nine-day period after the Winter Solstice, marking the beginning of the coldest days of the year. *A Collective Interpretation of the Seventy-two Pentads*, "At the

beginning of December, the weather is a little chill. In the middle of this month, the temperature drops drastically." People begin to feel cold at this point. As the saying goes, "During the Slight Cold and the Great Cold, people huddle against the coldness. Outside the window, the snow is falling heavily. While in the house, the stove fire is burning brightly."

In ancient China, the Slight Cold can be divided into three pentads (*hou*). When the Slight Cold comes, wild geese all move to north, magpies start to build their nest, and pheasants begin to tweet. The ancient Chinese people believed that the wild geese migrate owing to the *yin-yang* change. When *yangqi* is moving, the wild goose begin to move northward. At this time, magpies can be seen everywhere in the north. They sense *yangqi* and start to build nests. In addition, the pheasant would tweet for feeling the growth of *yangqi* at the time close to fourth nine-day period.

3. Customs of Slight Cold

【Diet】It is a good time to have spicy hot pot and stewed lamb in brown sauce on Slight Cold. The daily diet of residents centers on the food with warm nature, such as mutton and dog meat. Among the food, mutton is the most common one. Some restaurants even introduce the mutton soup with Chinese angelica and ginger. In recent years, some traditional mutton dishes in winter reappear on the dining table, reproducing the dining customs in cold days.

【Event】As the saying goes, "During the Slight Cold and Great Cold, people huddle against the coldness." The Slight Cold is around the third nine-day period after the Winter Solstice and it is a good time for people to do more physical exercises and improve their physical fitness. Southerners have a set of sports exercise method with regional characteristics, such as rope skipping, kicking shuttlecock and iron ring pushing. If it happens to snow, people would be in higher spirit. They can have a snowball fight or make a snowman. It is helpful for them to keep body warm and promote blood circulation.

Great Cold

1. Calculation Formula of Great Cold

The calculation formula: (Y×D+C)-L

The interpretation of formula: Y=the latter two digits of the year number, D=0.2422, L= the number of leap year, 21st C=20.12, 22nd C=20.84("C" stands for century.).

For example, the calculating date of Great Cold in 2089=(88×0.2422+20.12)-[(89-1)÷4]=41-22=19. Therefore, the day will fall on January 19th.

Exceptions: The date of Great Cold in 2082 will fall on the day after the calculating result.

2. Features of Great Cold

Great Cold (*Dahan*) is the last one of the Twenty-four Solar Terms in the traditional Chinese calendar and happens around January 20th in the solar calendar, when the sun reaches the celestial longitude of 300°. *San Li Yi Zong* quoted by *Shou Shi Tong Kao* explains, "Great Cold means the

cold weather reaches the extreme." The polar outbreak frequently invades the south. It is the coldest time of the year in most parts of China, with the strong wind, low temperature and the ground covered with accumulated snow, presenting a freezing cold winter world of ice and snow. During this period, the railway, post and telecommunications, petroleum, transmission line, waterage and other branches pay much attention to the precautions of the strong wind and the heavy snow.

3. Customs of Great Cold

【Diet】As the most important festival Spring Festival is approaching, Chinese people begin to be busy with removing the old ornaments, pickling the New Year's dishes and doing Spring Festival shopping. During this time, there is another important day for the northerners, Laba Festival. It falls on the 8th day of the 12th lunar month. People celebrate this day with Laba porridge, a delicious porridge mixed with grains, corn, chestnut, red date, lotus seed and other ingredients. Laba porridge originated from Buddhism. It is said that Shakyamuni obtained Buddha-hood on this day. Therefore, the monks cook porridge for the Buddha on this day, and this practice gradually becomes prevalent among common people.

In South China, common people are accustomed to cooking steamed glutinous rice in earthen pot when the Great Cold comes. The glutinous rice is sweet in taste and warm in nature, which has the effect of keeping from the cold and providing nourishment.

【Event】During the period from the Great Cold to the Start of

Spring, there are many important folk customs and festivals, such as Year-end Banquet (*weiya*), Kitchen God Worshipping and New Year's Eve. Sometimes even the largest festival—the Spring Festival is also among this solar term. *Weiya* pronounced in Chinese originates from the custom of worshipping the God of the Earth.

"Head tooth"(*touya*) happens on the 2nd of the 2nd month in lunar calendar and the *weiya* ends on 16th of the 12th lunar month. Common people celebrate these two days with spring pancake (which is called popiah in the south). In addition, *weiya* is also an activity for businessmen to reward the hard work of the past year. Sliced boiled chicken is an indispensable dish on the banquet.

During this period, people have much to do for Spring Festival in addition to farm work, such as going to market, buying goods, writing Spring Festival couplets, preparing various sacrifices, doing some cleaning and replacing the old with the new. At the same time, they offer sacrifices to their ancestors and various gods to pray for a good development in the coming year. Besides, the folks in ancient times in some areas would vie in purchasing sesame straw. Because of the old saying that "sesame stalks put forth flowers notch by notch", people sprinkle the sesame straw on the road for children to tread on New Year's Eve.

二月

二月有立春、雨水两个节气。

立春

1. 日期计算

计算公式：(Y×D+C)-L

公式解读：Y= 年数的后 2 位，D=0.2422，L= 闰年数，21 世纪 C=3.87，22 世纪 C=4.15，结果取整数。

举例说明：2017 年立春日期 =(17×0.2422+3.87)-(17÷4)= 3.9874，即 2 月 3 日。

2. 节气特点

立春是二十四节气中的第一个节气，为每年 2 月 3 日左右（农历正月初一前后），太阳到达黄经 315° 时。立春是从天文上来划分的，春代表温暖，鸟语花香；春代表生长，耕耘播种。据《月令七十二候集解》：“立春，正月节；立，建始也；五行之气往者过来者续于此；而春木之气始至，故谓之立也。”从立春当日一直到立夏前这段期间，都被称为春天。

立春之日迎春的习俗已有 3 000 多年历史，我国自官方到民间都对它极为重视。旧时，立春日天子亲率三公九卿、诸侯大夫去东郊迎春，祈求丰收。回来之后，要赏赐群臣，布德令以施惠兆民。后来

发展为全民参与的迎春活动。

3. 节气习俗

【饮食】取生菜、瓜果、饼糖，放入盘中为“春盘”，馈送亲友或自食。盘里主要有水果、蔬菜、糖果、饼、饵五种食物。蔬菜包括豆芽、萝卜、韭菜、菠菜、生菜、豆子、鸡蛋、土豆等。杜甫《立春》有言：“春日春盘细生菜，忽忆两京梅发时。”

立春之后的一段时间往往冷暖不定，人们杀菌并防寒，多于此时吃大蒜、洋葱、芹菜等“味冲”的食物，这对预防伤寒感冒等春季多发病症大有益处。

立春之后气候依然比较干燥，喝花茶可以帮助驱散冬季聚积在人体内的寒气和邪气。由于每一种花、草都有其相应的性、味、功效，如果使用得当，花草茶的确有一定的疗效，但需根据自身的体质进行

调整，饮用过量也会造成身体不适。

【活动】立春后气温回升，万物复苏，意味着新的一个轮回已开启，春耕大忙季节即将在全国范围陆续开始。但立春时节还属于农闲时节，不过此时家家都在“忙”，人们忙着除旧饰新，腌制年肴，准备年货，因为我国人最重要的节日——春节就要到了，此即立春迎年的风俗。立春祭是一项传统民俗文化活动，人们在黄道吉日举行祭祝祈年活动，以表达对神灵的崇拜、敬重和敬畏，并祈求丰收。立春祭活动内容包括祭春神、太岁、土地等众神，还有鞭春牛、迎春、探春、咬春等活动。立春岁首拜太岁是我国民间一种化煞消灾、祈福纳吉的古老传统习俗。

雨水

1. 日期计算

计算公式：$(Y\times D+C)-L$

公式解读：Y= 年数的后 2 位，D=0.2422，L= 闰年数，21 世纪 C=18.73，22 世纪 C=4.15，结果取整数。

举例说明：2008 年雨水日期 $=(8\times 0.2422+18.73)-[(8-1)\div 4]=20-1=19$，即 2 月 19 日。

例外：2026 年计算得出的雨水日期应调减一天为 18 日。

2. 节气特点

雨水是二十四节气中的第二个节气，时为每年正月十五前后（公历 2 月 19 ~ 21 日），太阳位于黄经 330°。太阳的直射点也由南半球逐渐向赤道靠近，这时的北半球，日照时长和强度都在增加，气

温回升较快，来自海洋的暖湿空气开始活跃，并渐渐向北挺进。与此同时，冷空气在减弱的趋势中并不甘示弱，与暖空气频繁较量，既不甘退出主导的地位，也不肯收去余寒。这导致降雨开始，雨量渐增。雨水之前天气寒冷，雨水之后气温一般可升至0℃以上，雪渐少而雨渐多。据《月令七十二候集解》："正月中，天一生水。春始属木，然生木者必水也，故立春后继之雨水。且东风既解冻，则散而为雨矣。"雨水节气前后，草木萌动，春天就要到了。

3. 节气习俗

【饮食】雨水节气天气虽然逐渐转暖，但早晚依旧寒冷，多风干燥，风邪会使人的皮肤、口舌干燥，嘴唇也会脱皮、干裂，因此应多食水果、蔬菜、西洋参和蜂蜜等以补充体内的水分和津液。春天万物生发，阳气上升，可多吃蛋白粉、莲子、百合、淮山药、薏米、绿豆、红枣和

枸杞子等。

【活动】撞拜寄，即为小儿认干亲，是雨水节气的一大习俗。在部分地区，雨水这一天的早晨，会有年轻妇女牵着幼小的儿女，在路边等待第一个从面前经过的行人。而一旦有人经过，也不管是男是女，是老是少，他们就拦住对方，让儿子或女儿跪地向他们磕头，认对方作干爹或干妈，希望孩子健康成长。

另外一大习俗就是占稻色。占稻色，就是通过爆炒糯谷米花，来占卜当年稻谷收获的丰歉。"成色"的好坏，就看爆出的糯米花多少，爆出来白花花的糯米越多，收成越好；而爆出来的米花越少，则意味着收成不好，米价将贵。

客家雨水节里，有个民俗是女儿给父母、女婿给岳父母送节。女婿送节的礼品通常是一条红棉带，称为"接寿"，祈求岳父母长命百岁。女儿则是炖了猪脚、鸡汤，用红纸、红绳封了罐口，由女婿给岳父母送去。这是女婿对辛苦将女儿养育成人的岳父母表示感恩。如果是新婚女婿送节，岳父母还要回赠雨伞，女婿出门奔波可以用来遮风挡雨，祝福女婿人生旅途顺利平安。

February

There are two solar terms in February—Start of Spring and Rain Water.

Start of Spring

1. Calculation Formula of Start of Spring

The calculation formula: (Y×D+C)-L

The interpretation of formula: Y = the latter two digits of the year number, D=0.2422, L=the number of leap year, 21st C=3.87, 22nd C=4.15("C" stands for century.). Finally, retrieve the integer.

For example, the date of the Start of Spring in 2017 = (17×.0.2422+ 3.87)-(17÷4)=3.9874, that is to say, the Start of Spring for 2017 would fall on February 3rd.

2. Features of Start of Spring

Start of Spring (*Lichun*) is the first among the Twenty-four Solar Terms and falls on around February 3rd (around the first day of the first month in the lunar calendar), when the sun reaches the celestial longitude of 315°. The Start of Spring is divided from the perspective of astronomy. Spring represents warmth and growth. In spring, the birds sing pleasantly and the flowers give forth fragrance. In the meantime, farmers also sow the seeds. *A Collective Interpretation of the Seventy-two Pentads* says, "*Lichun* is the solar term falling in the first month in Chinese lunar calendar. In Chinese, *li* means the beginning. For the five-elements motion, the last one is gone and the next one is coming repeatedly. At this time, the wood motion of spring appears, thus it is called *li* in Chinese." The spring lasts from the Start of Spring until the Start of Summer.

The custom of celebrating and greeting spring on the day of the Start of Spring has a history of more than 3 000 years. Chinese people have attached great importance to it from officials to civilians. In ancient times, on the Start of Spring, the emperor led three councilors, nine ministers,

the feudal princes and the bureaucrats to greet the spring on the eastern outskirts, praying for a good harvest. When the emperor came back, he would reward the officials and implement policy of benevolence to bestow favors on civilians. Later, it developed into a popular spring-greeting event.

3. Customs of Start of Spring

【Diet】Chinese people put various kinds of fruits, vegetables, candies and pies in one dish. This kind of dish is called "spring dish". People not only eat them with their own families, but also send them as gifts to their neighborhood. There are five primary ingredients—fruits, vegetables, candies, cakes and pastries. The vegetables mainly include bean sprouts, radishes, leeks, spinach, lettuce, beans, eggs and shredded potatoes. The poem *Lichun* written by Du Fu also mentions spring dish, "The fine lettuce in the spring dish makes me recall the blooming winter sweet in Chang'an and Luoyang."

During the period after the Start of Spring, the weather is sometimes cold and sometimes warm. If people want to sterilize and protect themselves from the coldness, they should increase the number of times they eat garlic, onions, celery and other strong-smelling food. That is beneficial for preventing from multiple respiratory infections such as catching a cold.

After the Start of Spring, it is still relatively dry. Drinking herbal tea can help disperse the cold and pathogenic *qi* that accumulate in the human body in winter. Since every kind of flower and herb has its corresponding property, taste and efficacy, herbal tea does have a certain health care effect if used properly. However, it needs to be adjusted according to different people's constitution. Excessive drinking can also cause physical discomfort.

【Event】After the Start of Spring, everything comes back to life due to the rise of temperature, which means that a new cycle has been opened. The season for spring ploughing is about to come in succession throughout the country. Although the beginning of spring season is still a slack season in agriculture, every family is pretty "busy" at this time. People are busy in removing old ornaments, pickling new dishes and buying new year's goods because the Spring Festival, the most important festival for Chinese people, is approaching. These activities are the traditional customs of welcoming Chinese New Year. Besides, worshipping god on the Start of Spring season is also a traditional folk cultural event. People will hold a festival to pray for a good harvest on an auspicious day and to express their worship, respect and awe of the gods. These activities include worshipping the god of the spring, of the land and so on. What's more, there are also other events like whipping the ox in spring, greeting spring, exploring spring and biting spring.

Rain Water

1. Calculation Formula of Rain Water

The calculation formula: (Y×D+C)-L

The interpretation of formula: Y = the latter two digits of the year number, D=0.2422, L=the number of leap year, 21st C=18.73, 22nd

C=4.15("C" stands for century.). Finally, round the result off to the nearest number.

For example, the date of Rain Water in 2008 = (8×0.2422+18.73)-[(8-1)÷4]=20-1=19, that is to say, the Rain Water for 2008 would fall on February 19th.

Exception: The date of Rain Water in 2026 will fall on the day before the calculating result, February 18th.

2. Features of Rain Water

Rain Water (*Yushui*) is the 2nd in the Twenty-four Solar Terms and falls on around the 15th of the first month of the year (vary from February 19th to February 21st in the solar calendar), when the sun reaches the celestial longitude of 330°. The subsolar point is also gradually approaching the equator from the Southern Hemisphere. Meanwhile, the duration and intensity of sunlight in the Northern Hemisphere are increasing. Therefore, the temperature rises rapidly. The warm and humid air from the ocean begins to become active and gradually advances northwards. At the same time, the cold air is also not willing to be in the weakening trend and fight against the warm air frequently. It is neither willing to withdraw from the dominant position, nor take away the remaining coldness. Therefore, the rainfall begins and the precipitation gradually increases. Before Rain Water, the weather is cold. In contrast, the temperature can generally rise above 0°C after Rain Water, with less snow and more rain. *A Collective Interpretation of the Seventy-two Pentads* says, "In the middle of first month, water flow out

of heaven. The spring is associated with wood. However, wood production can't exist without water. Hence Rain Water comes after the Start of Spring. Besides, the east wind can unfreeze the world and promote the appearance of rain." It means that everything begins to grow before and after Rain Water. Spring arrives.

3. Customs of Rain Water

【Diet】Although the weather in Rain Water is gradually warmer, it is still cold in the morning and evening. At the same time, it is windy and dry. The wind will make people's skin and tongue dry and cause the peeling of their lips. Therefore, more fruits, vegetables, ginseng, honey should be eaten to supplement the body's water and fluid. In the spring, everything grows and the *yangqi* rises. People can eat more protein powder, lotus seed, lily, Chinese yam rhizome, barley, mung bean, red date, medlar, etc.

【Event】A traditional custom is to find a godfather or godmother for little kids. In some areas, young women will hold their children and wait for the first pedestrian to pass by on the morning of Rain Water. Once someone passes by, they always stop this person and let their son or daughter kneel to him/her so as to recognize the person as godfather or godmother regardless of gender and age. The custom can be traced back to the health services of the past when many children's diseases could not be cured at all. Therefore, parents wanted a godfather or godmother to shower their blessings and good luck on their children.

Another custom is to predict the rice harvest according to the stir-

frying of glutinous rice. The quantity of glutinous rice can indicate the amount of puffed rice. More puffed glutinous rice stands for better harvest while the less puffed glutinous rice shows a poor harvest. The rice price will be more expensive accordingly.

For Hakka, one of the main customs during Rain Water is husbands visiting their parents-in-laws and giving gifts. According to the tradition, the gift is usually a red cotton belt from the son-in-law, which symbolizes good health and longevity. Another typical gift is a pot of cooked dish made by the daughters that contains pig's knuckles stewed with chicken soup, by which the son-in-law expresses his respect and gratitude. If the son-in-law is newly-married, the parents also need to give an umbrella in return. They wish the son-in-law a smooth and peaceful journey and the son-in-law can go out to keep from the wind and rain with this umbrella.

三月

三月有惊蛰、春分两个节气。

惊蛰

1. 日期计算

计算公式：(Y×D+C)-L

公式解读：Y= 年数的后 2 位，D=0.2422，L= 闰年数，21 世纪

C=3.87，22 世纪 C=5.63，结果取整数。

举例说明：2088 年惊蛰日期 = (88×0.2422+5.63)-(88÷4)=26-22 =4，即 3 月 4 日。

2. 节气特点

惊蛰是二十四节气中的第三个节气，时为公历 3 月 5 ~ 6 日之间，太阳到达黄经 345° 时。据《月令七十二候集解》："二月节……万物出乎震，震为雷，故曰惊蛰，是蛰虫惊而出走矣。"惊蛰也是春天的"复活节"，春芽和病毒同时复活。惊蛰前后乍寒乍暖，风的变化较大，气温明显回升，雨水增多。除东北、西北地区仍是银装素裹的冬日景象外，我国大部分地区平均气温已升到 0℃以上，华北地区日平均气温为 3 ~ 6℃，沿江江南地区为 8℃以上，而西南和华南已达 10 ~ 15℃。

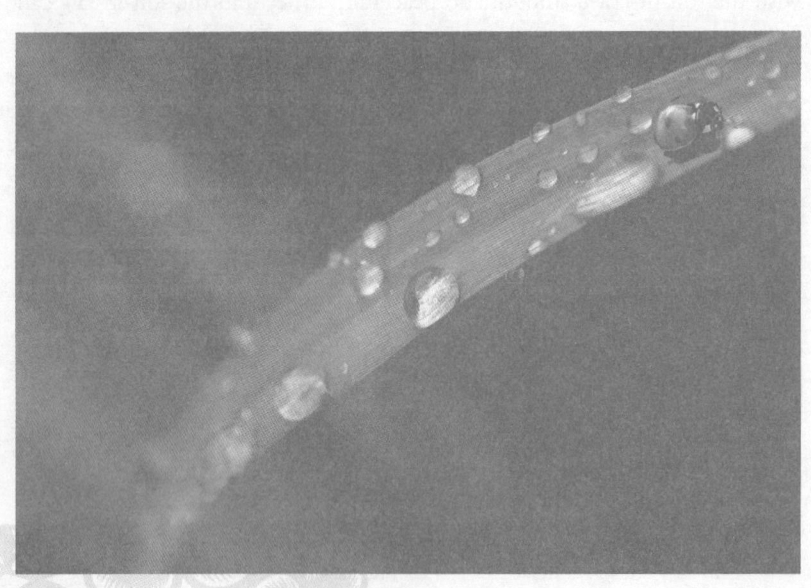

3. 节气习俗

【饮食】惊蛰一到天气回暖，清淡的食物有助于人体的新陈代谢。惊蛰时节气温还是普遍偏低，空气干燥，所以很容易导致上火，在日常的饮食上就要多增加生津润肺的食物，例如雪梨、银耳和春笋等。

由于惊蛰时节气温偏低，早晚温差大，所以这个节气还是很不适合吃偏冷的东西，入口的食物最好都是暖食。

民间素有“惊蛰吃梨”的习俗。在我国传统文化中，一般节日忌讳吃梨。不过惊蛰吃梨，寓意着和害虫分离，远离疾病。而且此时天气渐暖，空气干燥，吃梨可以润肺健脾，滋阴清热，有益于身体健康。梨可以生食，也可蒸、烤、榨汁或者煮水。

【活动】惊蛰到来后，虽然国内一些地方还难以见到花红柳绿的美景，但气温的升高已经能给人们一种“万物复苏”的感受。许多颇具象征意义的民俗亦应运而生。

有些地区在惊蛰节气会有蒙鼓皮的活动。惊蛰是雷声引起的。古人想象雷神是位鸟嘴人身、长了翅膀的大神，一手持锤，一手连击环绕周身的许多天鼓，发出隆隆的雷声。惊蛰这天，天庭有雷神击天鼓，人间也利用这个时机来蒙鼓皮，击鼓回应。

春雷也唤醒了蛰伏的小动物。过去卫生条件比较差，不少百姓家中就可能出现爬虫等。古时在惊蛰当日，人们会手持熏香、艾草，熏家中四角，以香味驱赶蛇、虫、蚊、鼠和霉味。

春分

1. 日期计算

计算公式：(Y×D+C)-L

公式解读：Y= 年数的后 2 位，D=0.2422，L= 闰年数，21 世纪 C=20.646，结果取整数

举例说明：2092 年春分日期 = (92×0.2422+20.646)-(92÷4)= 42-23=19，即 3 月 19 日。

例外：2084 年的计算结果加 1 日。

2. 节气特点

春分是二十四节气中的第四个节气，时为每年农历二月十五日前后（公历大约为 3 月 20 ～ 21 日期间），太阳位于黄经 0°（春分点）

时,是春季 90 天的中分点。春分这一天太阳直射地球赤道,南北半球季节相反,北半球是春分,在南半球来说就是秋分。春分是伊朗、土耳其、阿富汗、乌兹别克斯坦等国的新年,有着 3 000 年的历史。《月令七十二候集解》有言:“二月中,分者半也,此当九十日之半,故谓之分。”

春分节气有三个主要特点。一是昼夜平分,季节平分。春分的“分”有两个含义。一是指“昼夜平分”。春分之日太阳光直射在赤道上,几乎全球昼夜等长,即白天和黑夜的时间相等,都是 12 小时。此时北极点在经历了半年的黑夜后,在这一天初见阳光,而相对应的南极点则将在此日告别阳光。春分过后,阳光直射点位置便向北移,北半球白天时间开始变长,夜间时间变短了,所不同的是南半球则变得昼短夜长。二是指“季节平分”。若以立春至立夏这段时间作为春季,春分是春季的中分点,正好平分了春季。从立春到立夏正好 90 天,春分将其一分为二,因此得名“春分”。

二是春暖主导。就二十四节气所指的春季来看,春分前后往往有比较明显的天气变化。这个阶段的天气变化无常,冷暖交替明显。虽说冷空气还是常来常往,但已退居二线。来自海洋的暖湿空气,开始在我国占据主导地位。长江中下游、黄淮地区也都先后进入了气候学上所定义的春季,平均气温基本稳定在 10℃以上。此时需要警惕灾害性天气,北方会出现干旱、沙尘暴,南方则是低温冷害、倒春寒。

三是早稻插播,春季造林。春分时节我国大部分地区的越冬作物进入春季生长阶段。冬小麦开始返青拔节,农谚“春分麦起身”,预示着冬小麦进入了田间管理阶段。田里处处可以看到辛勤劳作的人

们。此时正是人们撒下种子，播种希望的时节。春分时节还是植树造林、移花接木的最佳时节，我国植树节定在了3月12日。有谚语曰："节令到春分，栽树要抓紧。春分栽不妥，再栽难成活。"可见节气对种树的重要性。

3. 节气习俗

【饮食】"春菜"是一种野苋菜，有些地方称之为"春碧蒿"。逢春分那天，春菜产地的人纷纷外出采摘春菜。在田野中搜寻时，春菜多为嫩绿色的细苗。采回的春菜一般会与鱼片"滚汤"，名曰"春汤"。

【活动】每年的春分，世界各地都会有数以千万计的人在做"竖蛋"试验。选择一个光滑匀称、刚生下来不久的新鲜鸡蛋，轻手轻脚地在桌子上把它竖起来。人们以此来庆祝春天的来临，春分成了竖蛋游戏的最佳时光，故有"春分到，蛋儿俏"的说法。

江南地区则流行犒劳耕牛、祭祀百鸟的习俗。春分一到，耕牛便开始一年的劳作，农民以糯米团喂耕牛表示犒赏，并祭祀百鸟，一则感谢它们提醒农时，二是希望鸟类不要啄食五谷，祈祷丰年。

春分这天农民会按习俗给自己放假，每家都要吃汤圆，而且还要把不用包馅的汤圆煮好，用细竹扦着置于室外田边地坎，名曰"粘雀子嘴"，免得雀子来破坏庄稼。

春天是孩子们放风筝的好时候，尤其是春分当天。甚至大人们也参与。风筝形状多样，有王字风筝、鲢鱼风筝和月儿光风筝等，其大者有两米高，小的也有六七十厘米高。市场上有卖风筝的，大部分是供孩子们玩耍的小型风筝。

March

There are two solar terms in March—Awakening of Insects and Spring Equinox.

Awakening of Insects

1. Calculation Formula of Awakening of Insects

The calculation formula: (Y×D+C)-L

The interpretation of formula: Y = the latter two digits of the year number, D = 0.2422, L= the number of leap year, 21st C=3.87, 22nd C=5.63("C" stands for century). Finally, round the result off to the nearest number.

For example, the date of Awakening of Insects in 2088 = (88×0.2422+ 5.63)-(88÷4)=26-22=4, that is to say, it will fall on March 4th.

2. Feature of Awakening of Insects

Awakening of Insects (*Jingzhe*) is the 3rd in the Twenty-four Solar Terms and begins between March 5th and March 6th of the solar calendar, when the sun reaches the celestial longitude of 345°. The ancient book *A Collective Interpretation of the Seventy-two Pentads* mentions:"In the second month ... all things are shocked, which is caused by thunder. Therefore, it is addressed as the Awakening of Insects. The insects come out because of the thunder." The Awakening of Insects is also "Easter" in spring. Spring buds and viruses are back to life at the same time. Before and after the

Awakening of Insects, the weather is sometimes cold and sometimes warm. The winds change quickly and the temperature rises markedly. What's more, the precipitation increases. With the exception of the northeastern and northwestern regions, which are still covered with heavy snow, the average temperature in most parts of China has risen to above 0°C. The average temperature varies from 3°C to 6°C in northern China and reaches over 8°C along the Yangtze River. In the southwestern and southern China, it has reached from 10°C to 15°C.

3. Customs of Awakening of Insects

【Diet】When the Awakening of Insects arrives, the weather gets warmer. Light food helps the body's metabolism. When the Awakening of Insects comes, the temperature is still generally low and dry. So it is easy to suffer from excessive internal heat. In the daily diet, there should be more food that can promote the secretion of saliva and moisten lung, such as snow pear, tremella, spring bamboo shoots and soon.

However, due to the big difference in temperature between morning and evening and the low temperature on Rain Water, it is still not suitable for people to eat cold food at this time. The warm food is a better choice.

Eating pears around the Awakening of Insects is a widely-practiced custom in China. In traditional Chinese culture, eating pears is a taboo on general holidays. However, eating pears on the Awakening of Insects means people can separate from pests and protect themselves from diseases. And as the weather gets warmer and the air becomes dry, people tend to feel their

mouths parched and tongues dry, which can cause colds or coughs. A pear is sweet, juicy and cold, moistening the lungs to arrest a cough. Therefore, pears are highly recommended during the Awakening of Insects. Pears can be eaten raw, steamed, extracted, baked or boiled.

【Event】When the Awakening of Insects arrives, the increase in temperature has given people the feeling that everything comes back to life although it is still difficult to see the beautiful scenery in some parts of the country. Accordingly, many symbolic folk customs have also emerged.

The activities of covering the drum head will be held in some areas. The Awakening of Insects is caused by the spring thunder. The ancient people imagined that the god of thunder was a winged God who had bird's beak and human's body. He carried a hammer and hit many drums that surrounded him. Thus, it thunders. On the Awakening of Insects, there is a thunder God hitting the drum in the heaven and the rest world also uses this opportunity to cover the drum and hit it to response.

Spring thunder also awakens dormant small animals. In the past, the sanitary conditions were relatively poor. Therefore, reptiles and ants might appear in common people's houses. In ancient times, people would fumigate the four corners of their houses with incense and wormwood to remove snakes, insects, mosquitoes, rats and mildew.

Spring Equinox

1. Calculation Formula of Spring Equinox

The calculation formula: (Y×D+C)-L

The interpretation of formula: Y = the latter two digits of the year number, D=0.2422, L=the number of leap year, 21st C=20.646("C" stands for century.). Finally, round the result off to the nearest number.

For example, the date of the Spring Equinox in 2092 = (92×0.2422+ 20.646)-(92÷4)=42-23=19, that is to say, it will fall on March 19th.

Exception: The date of the Spring Equinox in 2084 will be the day after the calculating result.

2. Features of Spring Equinox

Spring Equinox (*Chunfen*) is the 4th term in the Twenty-four Solar Terms and falls on around 15th in the 2nd lunar month (about March 20th or 21st in solar calendar), when the sun reaches the celestial longitude of 0°. It is the midpoint during the 90 days in spring. On the day of the Spring Equinox, the sun shines directly on the equator of the earth. Since the northern and southern hemispheres are opposite, the Northern Hemisphere is on the Spring Equinox while the Southern Hemisphere is on the Autumn Equinox. The Spring Equinox is the New Year in Iran, Turkey, Afghanistan, Uzbekistan and other countries. It has a history of 3 000 years. *A Collective Interpretation of the Seventy-two Pentads* says, "In the middle of the second month, it halves the 90 days. *Fen* in Chinese means the half. Therefore, the middle point of the 90 days can also be called *fen*."

The Spring Equinox has three main characteristics. To begin with, the day and night are equally divided and the season is equally divided. The *fen* of the Spring Equinox has two meanings. First, it means "the day and night are equally divided". On the day of the Spring Equinox, the sun shines directly on the equator. The day and night are almost the same in length in the world, that is, the time of day and night is equal, which is 12 hours. At this time, the North Pole will see the sun at the beginning of the day after experiencing 6 months of darkness. In contrast, the corresponding South Pole will bid farewell to the sun on this day. After the Spring Equinox, the subsolar point moves northwards. The Northern Hemisphere begins to have longer day and shorter night. The difference is that the Southern Hemisphere starts to have shorter day and longer night. The second refers to "the equally divided season". If the time from the Start of Spring to the Start of Summer is regarded as spring, the Spring Equinox is the mid-point of it, which is exactly the half of the spring. There are 90 days from the Start of Spring to the Start of Summer. The Spring Equinox divides it equally, hence it gains the name *Chunfen* in Chinese.

The second is spring warmth. In terms of the spring in the Twenty-four Solar Terms, there are more obvious weather changes around the Spring Equinox. The weather at this stage is changeable and the alternation of cold and warm is apparent. Although the cold air is often coming and going, it has taken a back seat. The warm and humid air from ocean begins to occupy a dominant position in our country. The middle and

lower reaches of the Yangtze River and the Huang-Huai region have also greeted the spring defined by climatology. The average temperature has basically stabilized above 10°C. At this time, people need to be vigilant for catastrophic weather. In the north, these are drought and sand storm. While in the south, the coldness becomes the main factor.

The third is early sowing and spring afforestation. On the Spring Equinox, overwintering crops in most parts of China enter the growth stage. Winter wheat begins to turn green. The peasant's proverb goes that "on the Spring Equinox wheat stands erect", which indicates that winter wheat has entered the field management stage. The hard-working people could be seen everywhere in the fields. This is the time when people sow seeds and sow hope. The Spring Equinox is also the best season for afforestation and transplanting trees. China's Tree-planting Day falls on March 12th. There is a saying, "When the Spring Equinox comes, the planting of trees must be implemented as soon as possible. If you cannot plant trees on this solar term, it would be difficult for trees to survive later." It can be seen that solar terms are important for planting trees.

3. Customs of Spring Equinox

【Diet】The "spring vegetable", usually called "artemisia" in some areas, is a kind of wild amaranth. On the day of the Spring Equinox, people will go to pick spring vegetable. When searching in the field, they often see green and thin artemisia. After returning home, they usually put the spring vegetable and fish fillets together and cook "hot soup", known as "spring

soup".

【Event】Standing an egg upright is a popular game across the world during the Spring Equinox. Choose a smooth, well-proportioned and fresh egg that has just been laid for a few days, then make it stand on end on the table lightly. People practice this tradition to celebrate the coming of spring. The Spring Equinox is the best time for egg balancing games, so there is a saying that "when the Spring Equinox arrives, the eggs are beautiful".

Rewarding cattle is popular in the southern area of the lower reaches of the Yangtze River. As the Spring Equinox comes, farm work starts and both the farmers and the cattle start to become busy. Farmers will reward cattle with sticky rice balls to express their gratefulness. Meanwhile, people will also make sacrifice to birds, to thank them for bringing signals for farm work and to ask them not to eat grains later in the year.

On the day of Spring Equinox, farmers are on holiday according to customs. Each family will eat sweet dumplings and they also have to cook dumplings that do not have sweet stuffing. These dumplings are placed on the edge of the outdoor field with bamboo branches. The aim is to seal sparrow's beak so as not to destroy the crops.

Spring is a good time for children to fly kites, especially on the day of the Spring Equinox. Even grown-ups also participate in this activity. The categories of kites include the kites with Chinese character *wang*, the squid kites, the moonlight kites and so on. The big ones are two meters high and the small ones are also sixty to seventy centimeters high. There are kites sold on the market, most of which are small and suitable for children to play.

四月

四月有清明、谷雨两个节气。

清明

1. 日期计算

计算公式：(Y×D+C)-L

公式解读：Y= 年数的后 2 位，D=0.2422，L= 闰年数，21 世纪 C=4.81，20 世纪 C=5.59。

举例说明：2088 年清明日期 =(88×0.2422+4.81)-(88÷4)=26-22=4，即 4 月 4 日。

2. 节气特点

清明是二十四节气中的第五个节气。清明象征着黄河流域万木凋零的寒冬过去了，风和日丽的春天来到了。《月令七十二候集解》有言："清明，三月节……物至此时皆以洁齐而清明矣。"清明一到，气温升高，生气始盛，大地呈现春和景明之象。这一时节万物"吐故纳新"，洁齐而清明。清明前后往往阴雨纷纷，惠风拂面，这时的风和雨都是人们喜欢的，"沾衣欲湿杏花雨，吹面不寒杨柳风"。清明节又叫踏青节，在仲春与暮春之交，也就是冬至后的第 104 天，是我国传统

节日之一，也是最重要的祭祀节日之一，是祭祖和扫墓的日子。汉族传统的清明节大约始于周代，距今已有 2 500 多年的历史。

1935 年，中华民国政府将 4 月 5 日定为国定假日清明节。2007 年 12 月 7 日，国务院第 198 次常务会议通过了修改《全国年节及纪念日放假办法》的决定，其中规定清明节放假一天。2008 年，清明节正式成为法定节假日。

3. 节气习俗

【饮食】清明正处于冷空气与暖空气交替之际，所以时常一会儿阳光灿烂，一会儿阴雨绵绵，人体常会出现不适，特推荐桑葚薏米炖白鸽作为食补佳品。饮食要清淡，多吃蔬菜、水果，以防上火。

清明前后，也不妨喝一些菊花茶。中医认为，菊花能疏风清热，有预防感冒、降低血压等作用。但菊花茶喝多也会伤肝，因此要适量

饮用。

【活动】荡秋千是我国古代清明节习俗。它的历史很古老，最早叫“千秋”，后改为“秋千”。古时的秋千多用树枝为架，再拴上彩带做成。后来逐步发展为用两根绳索加上踏板的秋千。荡秋千不仅可以增进健康，而且可以培养勇敢精神，至今仍为人们特别是儿童所喜爱。

鞠是一种皮球，球皮用皮革做成，球内用毛塞紧。蹴鞠，就是用足去踢球。这是古代清明节时人们喜爱的一种游戏。相传是黄帝发明的，最初目的是用来训练武士。

踏青又叫“春游”。古时叫“探春”“寻春”等。四月清明，春回大地，自然界到处呈现一派生机勃勃的景象，正是郊游的大好时光。

清明前后，春阳照临，春雨飞洒，种植树苗成活率高，成长快。自古以来，我国就有清明植树的习惯。清明节种树来源于丧葬习俗。早在西周时期，封建统治者就开始在坟前种树。据《礼记》中所述，孔子就曾在四方云游之前，为了将来能够确认辨认祖坟，就在其父母的坟头种植了松柏。不过，这个时候植树和清明节还并未被联系在一起，植树和清明节彻底融合是在汉朝。相传，西汉初期，汉高祖刘邦因为常年在外征战，没有时间回故乡祭祖，直到他做了皇帝后才有机会回家祭祖。但是因为常年未回乡，他没有找到父母的坟墓，后来在众臣的帮助下，他才在乱草丛中找到一块破旧的墓碑，于是便命人修坟立碑，并在坟前种植松柏作为标志。正好这一天是农历二十四节气中的清明，于是刘邦根据儒士的建议，将清明定为祭祖节。从此，每逢清明节，刘邦都会荣归故里，举行盛大的祭祖、植树活动。后来，这种习俗流传到民间，人们便将清明祭祖和植树结合在一起，并逐渐

形成一种固定的民俗。

谷雨

1. 日期计算

计算公式：(Y×D+C)-L

公式解读：Y= 年数的后 2 位，D=0.2422，L= 闰年数，21 世纪 C=20.1，20 世纪 C=20.888。

举例说明：2088 年谷雨日期 =(88×0.2422+20.1)-(88÷4)=41-22=19，即 4 月 19 日。

2. 节气特点

谷雨是二十四节气中的第六个节气，时为每年 4 月 19 日至 21 日太阳到达黄经 30° 时。据《月令七十二候集解》："三月中，自雨水后，土膏脉动，今又雨其谷于水也。"这时天气温和，雨水明显增多，对谷类作物的生长发育有很大影响。气象专家表示，谷雨是春季最后一个节气，谷雨节气的到来，意味着寒潮天气基本结束，气温回升加快，雨水丰沛，大大有利于谷类农作物的生长。这也是预防农作物遭受虫害的重要时期。

3. 节气习俗

【饮食】谷雨节气，自然界阳气上升快，容易引发人体内热而生"火"。此时常见的上呼吸道感染等疾病的特点是挟热、挟湿，如咽喉肿痛、痰多咳嗽、两眼发红、牙龈肿痛和口腔溃疡等。谷雨节气不宜食用太多温补性食物以免"助火"，应多食一些稍凉的食物，像黄瓜、凉瓜、山竹、马齿苋、芹菜、萝卜、绿豆和丝瓜等。

南方有谷雨摘茶的习俗，传说谷雨这天的茶喝了可以清火明目。所以，谷雨这天不管是什么天气，人们都会去山上摘一些新茶回来喝，以祈求健康。

【活动】谷雨以后气温升高，病虫害进入高繁衍期，为了减轻病虫害对作物及人的伤害，农民一边进田灭虫，一边张贴谷雨贴，进行驱凶纳吉的祈祷。

古时有"走谷雨"的风俗，谷雨这天人们走村串亲，或者到野外走走，寓意与自然相融合，强身健体。

April

There are two solar terms in April—Pure Brightness and Grain Rain.

Pure Brightness

1. Calculation Formula of Pure Brightness

The calculation formula: $(Y \times D+C)-L$

The interpretation of formula: Y = the latter two digits of the year number, D=0.2422, L=the number of leap year, 21st C=4.81, 20th C=5.59("C" stands for century.).

For example, the calculating date of the Pure Brightness in 2088 $=(88 \times 0.2422+4.81)-(88 \div 4)=26-22=4$, that is to say, the day will fall on April 4th.

2. Features of Pure Brightness

Pure Brightness (*Qingming)* is the 5th term in the Twenty-four Solar Terms. *A Collective Interpretation of the Seventy-two Pentads* says, "Pure Brightness begins in the third month of Chinese lunar calendar... During this period, everything appears to be pure and bright." When Pure Brightness turns up, the temperature rises and the world is full of vigour. The spring also fills the air with warmth. At this time of the year, everything is "new", showing the picture of pureness and brightness. Before and after Pure Brightness, it is often rainy and windy. The wind and rain of this time are all liked by people. As this poem goes, "In the season when apricot flowers bloom, the drizzle wets my gown; the wind that stirs willows fails to chill me." It falls at the turn of the mid-spring and late spring, namely, the 104th day after the Winter Solstice. The Pure Brightness is one of the traditional Chinese festivals and one of the most important sacrificial festivals. It is the day of ancestor worship and grave sweeping. The traditional Pure Brightness of the Chinese Han nationality began around the Zhou Dynasty and has a history of more than 2 500 years.

In 1935, the government of the Republic of China clearly stipulated that April 5th marks the national holiday the Pure Brightness. On December 7th, 2007, the decision to amend "the Stipulation Measures of National Festival and Memorial Holiday", was passed in the 198th executive meeting of the State Council, which ruled that all citizens can have a day off on the Pure Brightness. In 2008, The Pure Brightness officially became a legal holiday.

3. Customs of Pure Brightness

【Diet】The Pure Brightness falls at the turn of cold air and warm air. The weather is sunny for a while and rainy for a while. It is easy for people to feel sick. The stewed white pigeons with mulberry and glutinous rice can work as food supplements. The diet should be light. More vegetables and fruits can protect people from internal heat.

Before and after the Pure Brightness, you may as well drink chrysanthemum tea. Traditional Chinese medicine believes that chrysanthemum can disperse wind and heat, with the effect of preventing

from the cold, lower blood pressure and so on. However, chrysanthemum tea will also hurt the liver, so people had better drink in moderation.

【Event】Swing is the custom of ancient Pure Brightness. It has a long history. Originally, the swing was called *qianqiu*. And later it was changed into *qiuqian*. In ancient times, the swing was made of tree branches and then tied with ribbons. Later, it gradually developed into a swing with two ropes and pedals. Swing can not only improve health, but also can cultivate courage. Nowadays, people, especially the children, are still fond of paying on the swing.

Ju is a kind of leather ball. The ball skin is made of leather and the ball is filled with feather. *Cuju* is to play the ball with your feet. This was a popular game in ancient times during Pure Brightness. It is said that the Yellow Emperor invented it. The initial purpose was to train warriors.

Taqing is also called spring outing. In ancient times, it was addressed as exploring the spring and seeking the spring. In April, the spring is coming. The natural world shows a scene of vitality. It is a good time for outings.

Around the Pure Brightness, there are warm sun and plenty rain. The survival rate of planting trees is high. The trees can grow fast. Therefore, China has had the habit of planting trees on the Pure Brightness since ancient times. Planting trees on Qingming Festival stems from funeral customs. As early as the Western Zhou Dynasty, feudal rulers began to plant trees on the grave. According to *The Book of Rites*, the sage Confucius also planted pine and cypress on the graves of his parents so as to identify the ancestral graves in the future before travelling abroad. However, tree planting and Qingming Festival have not yet been linked at this time. It was in the Han Dynasty that people completely integrated tree planting with Qingming Festival. In the early Western Han Dynasty, Liu Bang, Emperor Gaozu of Han was said to have no time to return to his hometown to worship his ancestors because he was fighting battles all year round. When he ascended the throne, Liu Bang finally grasped a chance. However, he did not find his parents' graves owing to his departure from hometown for many years. Later, with the help of the ministers he found an old tombstone in the weeds. Then Liu Bang ordered his subordinates to build graves and plant cypresses in front of the graves as a sign. Coincidentally, this day is the Pure Brightness Day in the Twenty-four Solar Terms of the Chinese lunar calendar. At Confucian scholars' suggestion, Liu designated *Qingming* as the ancestor worship festival. Since then, he would return to his hometown every Qingming Festival, holding a grand ancestor worship and tree planting event. Afterwards, this custom spread far and wide among common people. Hence people combined ancestor worshipping on Qingming Festival and planting trees. It gradually developed into a fixed folk custom.

Grain Rain

1. Calculation Formula of Grain Rain

The calculation formula: (Y×D+C)-L

The interpretation of formula: Y = the latter two digits of the year number, D=0.2422, L=the number of leap year, 21st C=20.1, 20th C=20.888("C" stands for century.).

For example, the calculating date of Grain Rain in 2088 = (88×0.2422+20.1)-(88÷4)=41-22=19, that is to say, the day will fall on April 19th.

2. Features of Grain Rain

Grain Rain (*Guyu*) is the 6th term among the Twenty-four Solar Terms and falls on from April 19th to 21st each year, when the sun reaches the celestial longitude of 30°. *A Collective Interpretation of the Seventy-two Pentads* says, "In the middle of the third month, the soil becomes loose and soft since the Rain Water. Now, the rain begins to moisten grain." On the Grain Rain, the weather is mild and there is an obvious increase of rainfall, which has a great influence on growth of grain crops. Meteorological experts say that the Grain Rain is the last solar term in spring, and it signals the end of cold weather and a rapid rise in temperature and rainfall, which is extremely important for the growth of crops. It's also a key time to protect the crops from insect pests.

3. Customs of Grain Rain

【Diet】On the Grain Rain, *yangqi* in the natural world rises quickly, which is easy to induce the internal heat of human body. At this time, common diseases such as upper respiratory tract infections are characterized by heat and dampness, like sore throat, cough, red eyes, swollen gums, oral ulcers, etc. During the Grain Rain, people should not eat too much warm food to avoid "heat". In addition, they had better eat more food with cold nature, such as cucumbers, cool melons, mangosteen, purslane, celery, radish, mung beans and loofah.

There is an old custom in southern China that people drink tea on the day of Grain Rain. Spring tea during Grain Rain can help to remove heat from the body and is good for the eyes. Therefore, no matter what the weather is on this day, people will go to mountains to pick some fresh tea leaves to drink, so as to keep health.

【Event】As temperature rises, days after the Grain Rain enter a peak period of diseases and insect pests. In order to reduce damages to crops and people, farmers will kill insect pests in fields while post up Grain Rain posters to drive away evil force and pray for good luck.

This is an old custom on Grain Rain when people visit relatives, friends or neighbors, or go for a walk in the wild, to experience nature atmosphere and strengthen physical fitness.

五月

五月有立夏、小满两个节气。

立夏

1. 日期计算

计算公式：(Y×D+C)-L

公式解读：Y=年数后2位，D=0.2422，L=闰年数，21世纪C=5.52，20世纪C=6.318。

举例说明：2088年立夏日期=(88×0.2422+5.52)-(88÷4)=26-22=4，即5月4日。

例外：1911年的计算结果加1日。

2. 节气特点

立夏是农历二十四节气中的第七个节气，表示孟夏时节的正式开始。太阳到达黄经45°时为立夏节气。据《月令七十二候集解》："立夏，四月节……夏，假也。物至此时皆假大也。"立夏表示即将告别春天，是夏天的开始，也是温度明显升高，炎暑将临，雷雨增多，农作物进入生长旺季的一个重要节气。

立夏前后，虽气温快速回升，但我国只有福州到南岭一线以南地区真正进入夏季，北方部分地区这时则刚刚进入春季，全国大部分地区平均气温在 18 ~ 20℃。华南地区气温为 20℃左右。

立夏以后，江南正式进入雨季，雨量和雨天均明显增多，连绵的阴雨不仅导致作物的湿害，还会引起多种病害的流行。

3. 节气习俗

【饮食】立夏之后，天气逐渐转热，饮食宜清淡，应以易消化、富含维生素的食物为主，大鱼大肉和油腻辛辣的食物要少吃。

【活动】“夏”是“大”的意思，是指春天播种的植物已经长大了。古代，人们非常重视立夏的礼俗。在立夏这一天，古代帝王要率文武百官到京城南郊去迎夏，举行迎夏仪式，勉励农民抓紧耕种。据说，君臣一律穿朱色礼服，配朱色玉佩，连马匹、车旗都要朱红色的，以表达对丰收的企求和美好的愿望。

许多地方立夏还有尝新等节日活动，以祈健康好运。尝新的食物非常丰盛，有“九荤十三素”之说，九荤为鲫鱼、咸蛋、卤虾等，十三素包括樱桃、梅子、麦蚕、笋、蚕豆、豌豆、黄瓜、莴笋和萝卜等。

此外，“立夏吃蛋”的习俗由来已久。民间有俗语：“立夏吃个蛋，力气长一万。”从立夏这一天起，天气晴暖并渐渐炎热起来，许多人特别是小孩子会有身体疲劳、四肢无力的感觉，食欲减退，称之为“疰夏”。古人认为，鸡蛋圆圆的，象征生活圆满，立夏日吃蛋能祈祷夏季平安，经受住“疰夏”的考验。从营养学角度来看，鸡蛋是夏季进补佳品。立夏后，农事开始繁忙起来，人容易疲乏。吃鸡蛋可以快速补充体力，预防暑天常见的食欲不振、身倦肢软、消瘦等苦夏症状。

小满

1. 日期计算

计算公式：(Y×D+C)-L

公式解读：Y= 年数的后 2 位，D=0.2422，L= 闰年数，21 世纪 C=21.04，20 世纪 C=21.86。

举例说明：2088 年小满日期 =(88×0.2422+21.04)-(88÷4)=42-22=20，即 5 月 20 日或 21 日。

例外：2008 年的计算结果加 1 日。

2. 节气特点

小满是二十四节气中的第八个节气，时为每年 5 月 20 日到 22 日之间，太阳到达黄经 60° 时。《月令七十二候集解》有言："四月中，

小满者，物至于此小得盈满。”其含义是夏熟作物的籽粒开始灌浆饱满，但还未成熟。小满过后，天气逐渐炎热起来，雨水开始增多，预示着夏季闷热、潮湿的天气将要来临。

对于长江中下游地区来说，如果这个阶段雨水偏少，可能是太平洋上的副热带高压势力较弱，位置偏南，这就意味着到了黄梅时节，降水也会偏少。

南方地区该时节降雨多、雨量大。如果此时北方冷空气可以深入到我国较南的地区，南方暖湿气流也强盛的话，那么就很容易在华南一带造成暴雨或特大暴雨。因此，小满节气的后期往往是这些地区防汛的紧张阶段。

黄河中下游地区的小麦在此时刚刚进入乳熟阶段，非常容易遭受干热风的侵害，从而导致小麦灌浆不足、粒籽干瘪而减产。防御干热风的方法很多，比如营造防护林带、喷洒化学药物等，都是十分有效的措施。

3. 节气习俗

【饮食】小满时节，吃玉米、高粱、苡仁、扁豆、水芹、冬瓜、洋葱、马齿苋、鲫鱼和鲍鱼等食物，可以祛湿。

【活动】祭车神是一些农村地区古老的小满习俗。在相关的传说里，车神是一条白龙。在小满时节，人们在水车上放上鱼肉和香烛等物品祭拜，最有趣的地方是，在祭品中会有一杯白水，祭拜时将白水泼入田中，有祝水源涌旺之意。

相传小满为蚕神诞辰。古时，养蚕是江南农家的传统副业，因此江浙一带在小满节气期间有一个祈蚕节。丝绸业的商人会以丝业会馆或丝业公所的名义摆供祭神、演戏酬神，以此祈佑新丝上市、生

丝交易旺季的到来。蚕是娇养的“宠物”,很难养活。气温、湿度,桑叶的冷、熟、干、湿均影响蚕的生存。由于蚕难养,古代把蚕视作“天物”。为了祈求“天物”的宽恕和养蚕有个好的收成,因此人们在四月放蚕时节举行祈蚕节。

May

There are two solar terms in May—Start of Summer and Grain Buds.

Start of Summer

1. Calculation Formula of Start of Summer

The calculation formula: (Y×D+C)-L

The interpretation of formula: Y = the latter two digits of the year number, D = 0.2422, L= the number of leap year, 21st C=5.52, 20th C=6.318("C" stands for century.).

For example, the calculating date of the Start of Summer in 2088 = (88×0.2422+5.52)-(88÷4)=26-22=4, that is to say, the day will fall on May 4th.

Exception: The date of the Start of Summer in 1911 will be the day after the calculating result.

2. Features of Start of Summer

Start of Summer (*Lixia*) is the 7th term among the Twenty-four Solar Terms and indicates the formal start of the early summer. On this day, the sun's ray reaches an angle of 45° to the earth. *A Collective Interpretation of the Seventy-two Pentads* says, "The Start of Summer (*Lixia*) falls on the fourth month of the Chinese lunar calendar. The meaning of *li* is explained in the part of the Start of Spring. *Xia* represents growth. At this time, everything grows large." The Start of Summer signifies that people would soon bid farewell to spring. People customarily believe that the temperature rises obviously on the Start of Summer. The hot summer is approaching. In addition, there are more thunderstorms and crops will enter the peak season to grow.

Though the temperature will rise quickly during this period, only the southern areas of Fuzhou to Nanling has truly entered summer, yet in northern China spring just starts. The average temperature in most parts of the country is about 18°C to 20°C. The temperature in the South China is about 20°C.

After the day of Start of Summer, the south of Yangtze River formally enters the rainy season. The daily precipitation and rainfall increase significantly. What's more, the incessant rain not only causes wet damage to the crops, but also causes the spread of many diseases.

3. Customs of Start of Summer

【Diet】After the day of the Start of Summer, the weather gradually turns hot, and people had better eat light food. The diet should center on digestible food which is rich in vitamin. Rich food as well as greasy and spicy food should be eaten less.

【Event】*Xia* in Chinese means "big", expressing that the plants planted in spring have grown upright. In ancient times, people attached great importance to the customs of the Start of Summer. On the day of the Start of Summer, ancient Chinese emperors would lead their bureaucrats to the southern suburbs of the capital to stage a welcome rite for the Start of Summer and encourage their residents to seize the key time to do farm work. It's said that most of the decorations were arranged in the color red, including the emperor and the officials' dress, the jade pendants, the horses and the flags, in order to show respect to the god of summer and pray for a good harvest.

The custom of "tasting seasonal food" on the day of the Start of Summer prevails in many places of China. People act on the custom and tuck in seasonal food on the day to pray for good health and good luck. A wide choice of seasonal fruits and vegetables which include "nine kinds of meat and thirteen kinds of vegetables", such as carps, salted eggs, salted shrimp, cherries, green plums, bamboo shoots, wheat strips, broad beans, peas, cucumbers, lettuce, radish, etc.

In addition, the custom of "eating eggs on the Start of Summer" has a long history. Among folks, there goes a jingle, "whoever eats eggs on the day of the Start of Summer will have the strength to crush stones", to the effect that if you eat eggs on the day of the Start of Summer, you will be full of vim and vigor. From the Start of Summer, the weather is warm and gradually turns hot. Many people, especially children, will feel physically

tired and have a poor appetite, which is called "summer disease". In ancient China, people believed a round egg symbolized a happy life and eating eggs on the day of the Start of Summer was a prayer for good health. From the perspective of modern nutriology, the egg is the food of choice in summer to replenish the nutrition our body needs. In summer, people are busy in farm work and lose a lot of physical strength. So, at the time, eating eggs can not only quickly replenish the energy but also improve the resistance to diseases, such as loss of appetite, lassitude of limbs, loss of weight, etc.

Grain Buds

1. Calculation Formula of Grain Buds

The calculation Formula: (Y×D+C)-L

The interpretation of formula: Y = the latter two digits of the year number, D=0.2422, L=the number of leap year, 21st C=21.04, 22nd C=21.86("C" stands for century.).

For example, the calculating date of Grain Buds in 2088 =(88×0.2422+21.04)-(88÷4)=42-22=20, that is to say, the day will fall on May 20th or May 21st.

Exception: The date of Grain Buds in 2008 would be the day after the calculating result.

2. Features of Grain Buds

Grain Buds (*Xiaoman*), which falls on between May 20th and 22nd

each year, is the 8th term among the Twenty-four Solar Terms. On the day, the sun reaches the celestial longitude of 60°. *A Collective Interpretation of the Seventy-two Pentads* mentions, "(*Xiaoman*) falls in the middle of the 4th month. It means the grains have buds but are not fully ripe." After the Grain Buds, the weather gradually becomes hot and the rainfall begins to increase, indicating that the hot and humid weather of summer would come.

For the middle and lower reaches of the Yangtze River, the subtropical high-pressure forces in the Pacific may be weak if there is less rain at this stage. In addition, it is in southerly direction, which means that by the time of the Huangmei season, precipitation may also be less.

In the southern region, there is much and heavy rainfall. If the cold air in the north can go deep into the south of our country and the warm and humid air in the south also keeps strong, it would be easy to cause heavy rain or extraordinary rain storm in southern China. Therefore, the later period of Grain Buds is often the tense stage of flood prevention in these areas.

The wheat in the middle and lower reaches of the Yellow River has just entered the milky maturity at this time and is very vulnerable to the invasion of dry hot winds, resulting in insufficient wheat grouting and dry seed production. There are many ways to prevent dry hot winds, such as building forest shelter belts and spraying chemical drugs, which are very effective measures.

3. Customs of Grain Buds

【Diet】On the day of Grain Buds, it is appropriate for people to eat corn, sorghum, coix, lentils, parsley, melon, onion, purslane, carp and abalone to clear damp.

【Event】In some rural areas, offering sacrifice to waterwheel deity is an ancient traditional custom. According to legends, the deity of waterwheel is a white dragon. People place fish and meat and scented candles in front of the waterwheel. One special thing among the offerings is a glass of water. They spill it on farmlands during the ritual, wishing for watery and good harvest.

According to legends, Grain Buds is the birthday of silkworm deity. In ancient times, lots of people from the southern regions of the downstream Yangtze River reared silkworms to make a living. Therefore, people in Jiangsu and Zhejiang provinces will worship the silkworm deity on the day of Grain Buds. Those who engage in the silk industry thank the deity by offering sacrifices and staging performances. They pray to the deity for blessings and a boom season for the silk business. Silkworms were quite difficult to breed and people regarded them as the God-given gifts in ancient times. Temperature, humidity and the condition of mulberry leaves all affect the survival of silkworms. In order to pray for blessings and bloom in silk industry, people hold ceremonies to thank and celebrate the birthday of the deity in every April.

六月

六月有芒种、夏至两个节气。

芒种

1. 日期计算

计算公式：(Y×D+C)-L

公式解读：Y= 年数的后 2 位，D=0.2422，L= 闰年数，21 世纪 C=5.678，20 世纪 C=6.5。

举例说明：2088 年芒种日期 =(88 × 0.2422+5.678)−(88 ÷ 4)=

26-22=4，即 6 月 4 日。

例外：1902 年的计算结果加 1 日。

2. 节气特点

芒种是二十四节气中的第九个节气，时为公历每年 6 月 6 日前后，太阳到达黄经 75° 时。《月令七十二候集解》将芒种分为三候："一候螳螂生，二候鹏始鸣，三候反舌无声。"此时我国长江中下游地区进入多雨的黄梅时节。

这个节气气候复杂多变。气温攀升显著，雨量充沛，常见的天气灾害有龙卷风、冰雹、大风、暴雨和干旱等。由于天气炎热，已经进入典型的夏季，农事种作都以这一时节为界，过了这一节气，农作物的成活率就越来越低。

芒种时节沿江多雨，黄淮平原也即将进入雨季，此时，西南西部的高原地区冰雹天气开始增多。芒种前后若遇连阴雨天气及风雹等，往往使小麦不能及时收割、脱粒和贮藏而导致麦株倒伏、落粒，使眼看到手的庄稼毁于一旦。6 月份，无论是南方还是北方，都有可能出现 35℃以上的高温天气，黄淮地区、西北地区东部可能出现 40℃以上的高温天气，但一般不是持续性的高温。在华南的台湾、海南、福建、两广等地，6 月的平均气温都在 28℃左右，在雷雨之前，空气湿度大，确实是又闷又热。芒种期间，依然会有大暴雨。正常情况下，一般先进入梅雨期的是湖南、江西中部、浙江南部地区，入梅后如同华南一样，该地区的主汛期开始，时有暴雨发生，山区地区需要警惕局地大暴雨引发的山洪、泥石流等次生灾害。

3. 节气习俗

【饮食】我国有些地方有谚语说："芒种夏至天，走路要人牵；牵

的要人拉,拉的要人推。”这形象地表现了人们在这个时节的懒散。医生提醒,人们要使自己的精神保持轻松、愉快的状态。夏日昼长夜短,午休可助缓解疲劳,有利于健康。芒种时气候开始炎热,是消耗体力较多的季节,要注意补充水分,多喝水。

饮食调养方面,唐朝的孙思邈提倡人们“常宜轻清甜淡之物,大小麦曲,粳米为佳”,又说“善养生者常须少食肉,多食饭”。在强调饮食清补的同时,他告诫人们食勿过咸、过甜。在夏季,人体新陈代谢旺盛,汗易外泄,耗气伤津,故宜多吃水果蔬菜。

【活动】芒种时节送花神是一种古老的民间祭祀习俗。人们认为,芒种已过,百花开始凋零,花神退位,故民间多在芒种日举行祭祀花神的仪式,饯送花神归位,同时表达对花神的感激之情,盼望来年再次相会。

每年五六月份也是南方梅子成熟的季节,人们会在芒种前后煮梅。这一习俗历史悠久,早在夏朝便有了。三国时有“青梅煮酒论英雄”。青梅含有多种天然优质有机酸和丰富的矿物质,具有净血、整肠、降血脂、消除疲劳、美容、增强人体免疫力等独特功能。但是,新鲜梅子大多味道酸涩,难以直接入口,需加工后方可食用,这种加工过程便是煮梅。

夏至

1. 日期计算

计算公式:(Y×D+C)-L

公式解读:Y= 年数的后 2 位,D=0.2422,L= 闰年数,21 世纪

C=21.37，20 世纪 C=22.20。

举例说明：2088 年夏至日期 =(88×0.2422+21.37)-(88÷4)=42-22=20，即 6 月 20 日。

例外：1928 年的计算结果加 1 日。

2. 节气特点

夏至是二十四节气中的第十个节气，时为每年公历 6 月 21 日或 22 日。夏至这天，太阳直射地面的位置到达一年的最北端，几乎直射北回归线。此时，北半球的白昼达最长。我国民间把夏至后的 15 天分成三“时”，一般头时 3 天，中时 5 天，末时 7 天。这期间我国大部分地区气温较高，日照充足，作物生长很快。据《月令七十二候集解》：“夏至，五月中……万物于此皆假大而至极也。”此时的降水对农业产量影响很大，有“夏至雨点值千金”之说。

夏至时节正是江淮一带的“梅雨”季节，空气非常潮湿，冷、暖空气团在这里交汇，并形成一道低压槽，导致阴雨连绵的天气。

夏至以后地面受热强烈，空气对流旺盛，午后至傍晚常易形成雷阵雨。这种热雷雨骤来疾去，降雨范围小。夏至和冬至一样，都是反映四季更替的节气。天文学上规定，夏至为北半球夏季的开始。夏至过后，太阳直射点开始由北回归线逐渐向南移动，北半球白昼逐日变短，黑夜逐日变长。

在大多数情况下，夏至正值长江中下游、江淮流域梅雨季，暴雨天气频频出现，容易形成洪涝灾害，甚至对人们的生命财产造成威胁，应注意加强防汛工作。

3. 节气习俗

【饮食】中医认为凡苦味蔬菜大多具有清热解毒的功效，所以苦瓜就成了夏至的最佳蔬菜。人们用苦瓜做汤、做菜、做饮品。苦瓜熟食性温，生食性寒，脾虚胃寒者不要多吃，也不要生吃，可以与热性食物搭配食用，如与辣椒一起清炒。

民间向来有“冬至饺子夏至面”之说。夏至这天，多地有吃面习俗，据说这一天吃一碗面，可祛除人体内滞留的潮气和暑气，也有为家人祈求身体康健、避恶之意。

夏至后，气温逐渐升高，人体出汗量也会随之增加，因此人体需水量大。对此，还可以在饮食上加以调节，如喝些绿豆汤、淡盐水等。需注意的是，绿豆汤不要多喝，更不能当水喝。属于寒凉体质和体质虚弱之人也不适宜饮用绿豆汤。

【活动】有些地方在夏至时兴称重量。据说在这一天称了体重后，高温酷暑都不怕。古时候缺医少药，一旦生病就难以很快痊愈，人们对夏至称重情有独钟。这也寄托了人们希望自己健康长寿的美好愿望。

June

There are two solar terms in June—Grain in Ear and Summer Solstice.

Grain in Ear

1. Calculation Formula of Grain in Ear

The calculation formula: (Y×D+C)-L

The interpretation of formula: Y = the latter two digits of the year number, D = 0.2422, L= the number of leap year, 21st C=5.678, 22nd C=6.5 ("C" stands for century.).

For example, the calculating date of Grain in Ear in 2088= (88×0.2422 +5.678)-(88÷4)=26-22=4,that is to say, the day will fall on June 4th.

Exception: The date of Grain in Ear in 1902 would be the day after the calculating result.

2. Features of Grain in Ear

Grain in Ear (*Mangzhong*) is the 9th term among the Twenty-four Solar Terms and begins around June 6th of the solar calendar, when the sun reaches the celestial longitude of 75°. In *A Collective Interpretation of the Seventy-two Pentads*, the Grain in Ear was divided into three pentads: "The mantis comes out for one pentad, the shrike starts to sing and the mockingbird stops tweet." During this period, areas around the middle stream and downstream of the Yangtze River enter the rainy season.

The climate is complex and changeable during this solar term. The

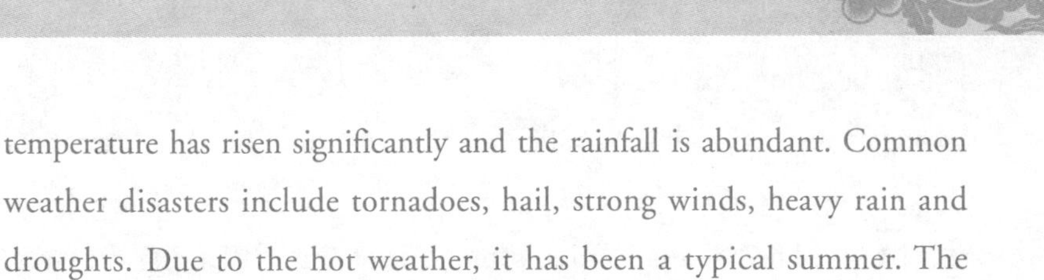

temperature has risen significantly and the rainfall is abundant. Common weather disasters include tornadoes, hail, strong winds, heavy rain and droughts. Due to the hot weather, it has been a typical summer. The agricultural farming is all bounded by this season. After this solar term, the survival rate of crops is getting lower and lower.

During the Grain in Ear, it is rainy in the regions along the river and the Huanghuai Plain is about to enter the rainy season. At this time, the hail weather in the western plateau of the southwest will begin to increase. If rain, wind and hail strike before and after Grain in Ear, wheat can not be harvested, threshed and stored in time, causing wheat to fall. Thus, the crops are destroyed. In June, whether it is in the south or the north, there is a possibility of scorching weather of above 35°C. High-temperature weather of above 40°C may occur in the Huanghuai region and the eastern part of the northwest region, but it is generally not a continuous high temperature. In southern China like Taiwan, Hainan, Fujian, Guangdong and Guangxi, the average temperature in June is around 28°C. Before a thunderstorm comes, the air is humid. It is indeed sultry and hot. The aim is to remind people to prevent heatstroke, air conditioning disease and acute gastroenteritis. During Grain in Ear, there will still be heavy rain. Under normal circumstances, the first batch of regions to enter the Meiyu period are Hunan, central Jiangxi and southern Zhejiang. After entering Meiyu period, the main flood season in the area begins as in southern China, and there is heavy rain. The people in mountainous areas need to be alert to

secondary disasters such as flash floods and debris flow caused by heavy rain.

3. Customs of Grain in Ear

【Diet】There is a proverb in some parts of China, "During the Grain in Ear, you need someone to hold you when walking; the man who holds you needs someone to drag; the man who drags needs someone to push." That shows the image of people's laziness at this time of the year. People should keep their minds relaxed and be cheerful, the doctor warns. In summer, the day is long and the night is short. Noon breaks can help relieve fatigue, which is good for health. In addition, the weather is getting hot. It is in this season that people exert much physical strength. Therefore, people have to drink more water to rehydrate the body.

In terms of diet, people should "eat light food during the Grain in Ear, such as wheat koji and rice", that's the healthcare advice given more than a thousand years ago by Chinese pharmaceutical expert Sun Simiao of the Tang Dynasty. Generally, during this season, people are encouraged to consume less meat but more rice and grains in order to keep healthy. In addition to the diet supplements, people are also warned not to eat too salty and too sweet food. In summer, the body's metabolism remains high. The weather is hot during the Grain in Ear period, therefore vegetables and fruits of a cool nature are recommended.

【Event】As the flowers become withered and fallen since the Grain in Ear, the ancient people in China held sacrificial ceremonies to say farewell to the flora and showed their gratitude and their eagerness to see the flowers

again next year. The flora will go back to heaven and come back to earth next year.

The harvest season for plums is from late May to early June in South China. Hence, it is a tradition to boil plums around the *Mangzhong* period, which can be traced back to the Xia Dynasty. There was an allusion that Cao Cao and Liu Bei, two central figures in the Three Kingdoms period, talked about heroes while boiling green plums. Green plums contain a variety of natural and high-quality organic acids and are rich in minerals. They can help clean blood, lower blood lipids, eliminate tiredness, improve one's looks and develop immunity. However, fresh plums taste so sour that greatly affect mouthfeel. Thus, it is accustomed to cooking the fruits before being eaten. This processing procedure is called boiling plums.

Summer Solstice

1. Calculation Formula of Summer Solstice

The calculation formula: (Y×D+C)-L

The interpretation of formula: Y=the latter two digits of the year number, D=0.2422, L=the number of leap year, 21st C=21.37, 20th C=22.20 ("C" stands for century.).

For example, the calculating date of the Summer Solstice in 2088 = (88×0.2422+21.37)-(88÷4)=42-22=20, that is to say, the day will fall on June 6th.

Exception: The date of the Summer Solstice in 1928 would be the day

after the calculating result.

2. Features of Summer Solstice

Summer Solstice (*Xiazhi*), beginning around June 21st or June 22nd, is the 10th solar term among the Twenty-four Solar Terms. On this day, the position of the sun's ray directly on the ground reaches the northernmost point of the year, where is the Tropic of Cancer. And it is also when the Northern Hemisphere sees the longest day and the shortest night of the year. People in China divide the 15 days after the Summer Solstice into 3 *shi*. Generally, the first *shi* is 3 days and the middle *shi* lasts 5 days. The last *shi* is 7 days. During this period, in most regions of China the temperature is high. In addition, the crops grow rapidly with abundant sunshine. *A Collective Interpretation of the Seventy-two Pentads* says, "Summer Solstice lies in the middle of the fifth month...Everything is ripe at this time." At this time, precipitation has a great influence on agricultural production. There is a saying that "the rain is worth thousands of dollars on the Summer Solstice".

When the Summer Solstice comes, it is the plum rainy season in the Jianghuai Region. The air is humid and relatively cold, and when broached with warm air, a trough of low pressure takes form and leads to continuous rainy days.

After the day of Summer Solstice, affected by the strong heat exposure of the ground, air convection is active. Thunder showers occur easily after noon and at sunset. This kind of hot thunder shower comes and goes quickly. Like the Winter Solstice, the Summer Solstice is a solar term that

reflects the change of the four seasons. In astronomy, the Summer Solstice is the beginning of the summer in the Northern Hemisphere. After the Summer Solstice, the subsolar point begins to move southwards from the Tropic of Cancer and the day is gradually shorter and the night becomes longer in the Northern Hemisphere.

In the Summer Solstice period, the plum rains are still dominant with occasional rainstorms in the middle and lower reaches of the Yangtze River and the Jianghuai River. Meanwhile, the continuous rain could easily lead to floods and even threaten people's safety and their property. People should pay attention to strengthening flood control.

3. Customs of Summer Solstice

【Diet】Traditional Chinese medicine believes that bitter food has heat-clearing and detoxifying effect. Among them, bitter gourd is considered the best choice. Chinese people cook it by soup, drinks and fried dishes during the Summer Solstice. The bitter gourd is warm in nature if cooked and cold in nature if eaten raw. Those with a weak spleen and cold stomach should not eat more, nor do they eat raw. They can eat bitter gourd with hot food, such as stir-fried bitter melon with pepper.

There is a saying: "Eat dumplings on the Winter Solstice and eat noodles on Summer Solstice." In many parts of China, people have noodles on that day. It is said that eating hot noodles can help ward off evils and summer heat inside human bodies. People eat a bowl of noodles on the day to pray for good health and good luck.

After the Summer Solstice, the temperature gradually goes up, and the amount of sweating in the human body will also increase. Therefore, the human body needs a large amount of water. In this regard, people can also balance their diet, such as drinking mung bean soup, brackish water and so on. It should be noted that mung bean soup should not be drunk more. In addition, mung bean soup is not suitable for people who have cold and weak constitution.

【Event】In some places, the custom of weighing people at the Summer Solstice is still prevalent today. It was believed this activity can bring health and good luck to the people weighted. They will stay healthy in the hot summer. In ancient times, there was a lack of doctors and medicine. Once someone was sick, it was difficult to recover. Therefore, people had a special love for weighing on the Summer Solstice. This also shows people's wish for a long and healthy life.

七月

七月有小暑、大暑两个节气。

小暑

1. 日期计算

计算公式：(Y×D+C)-L

公式解读：Y= 年数的后 2 位，D=0.2422，L= 闰年数，21 世纪 C=7.108，20 世纪 C=7.928。

举例说明：2088 年小暑日期 =(88×0.2422+7.108)-(88÷4)=28-22=6，即 7 月 6 日。

例外：1925 年和 2016 年的计算结果加 1 日。

2. 节气特点

小暑是二十四节气中的第十一个节气，时为每年 7 月 7 日或 8 日太阳到达黄经 105°时。据《月令七十二候集解》："六月节。暑，热也。就热之中分为大小，月初为小，月中为大，今则热气犹小也。"意思是小暑为"小"热，还不十分热。全国大部分地区基本符合。全国的农作物都进入了茁壮成长阶段，需加强田间管理。

南方地区小暑时平均气温为 33℃左右。7 月中旬，华南东南低

海拔河谷地区，可能开始出现日平均气温高于30℃、日最高气温高于35℃的情况。在西北高原北部，此时仍可见霜雪，相当于华南初春时节的景象。

小暑开始，江淮流域梅雨先后结束，东部淮河、秦岭一线以北的广大地区开始了来自太平洋的东南季风雨季，降水明显增加，且雨量比较集中；华南、西南、青藏高原也处于来自印度洋和我国南海的西南季风雨季中；而长江中下游地区则一般为副热带高压控制下的高温少雨天气。有的年份，小暑前后北方冷空气势力仍较强，在长江中下游地区与南方暖空气势均力敌，出现锋面雷雨。小暑时节的雷雨常是“倒黄梅”的天气信息，预兆雨带还会在长江中下游维持一段时间。

小暑前后，我国南方大部分地区各地进入雷暴最多的季节，某些年份也曾干旱。雷暴是一种剧烈的天气现象，常与大风、暴雨相伴出现，有时还有冰雹，容易造成灾害。华南东部，小暑以后因常受副热带高压控制，多连晴高温天气，开始进入伏旱期。因此，小暑时期还要做好防汛和抗旱工作。

3. 节气习俗

【饮食】一直以来，民间素有小暑吃藕的习俗，藕中含有大量的碳水化合物及丰富的钙磷铁等和多种维生素，适合夏天食用。鲜藕以小火煨烂，切片后加适量蜂蜜，有安神入睡之功效。

俗话说：“小暑黄鳝赛人参。”黄鳝生于稻田、小河、池塘、湖泊等淤泥质水底层，以小暑前后一个月的夏鳝鱼最为滋补味美。黄鳝蛋白质含量较高，并含有多种矿物质和维生素，铁的含量比鲤鱼、黄鱼等高一倍以上，还可降低血液中胆固醇的浓度，防治动脉硬化引起的

心血管疾病。

“热在三伏”，小暑是进入伏天的开始，天气热的时候要多喝粥，可用荷叶、土茯苓、扁豆、薏米、猪苓、泽泻、木棉花等材料煲成的消暑汤或粥，或甜或咸，非常适合此节气食用。

北方会在小暑、大暑期间喝羊汤，据说可防疾病。第一可以滋补身体；第二“羊”与“阳”谐音，古人认为夏季阳气丧失较多，这样能够增加阳气。

【活动】小暑时节，民间有晒书画、衣服的习俗。民谚有云：“六月六，人晒衣裳龙晒袍”，“六月六，家家晒红绿”。“红绿”就是指五颜六色的各种衣服。因为这一天（基本在小暑前夕）几乎是一年中气温最高、日照时间最长、阳光辐射最强的日子，所以家家户户多会不约而同地选择这一天“晒伏”，把存放在箱柜里的衣服放到外面接受阳光的暴晒。

“食新”即在小暑过后尝新米。农民将新割的稻谷碾成米后，做成饭祭祀五谷大神和祖先，表示对大自然以及祖先的感恩，然后人人吃“尝新酒”。

大暑

1. 日期计算

计算公式：(Y×D+C)-L

公式解读：Y= 年数的后 2 位，D=0.2422，L= 闰年数，21 世纪 C=22.83，20 世纪 C=23.65。

举例说明：2088 年大暑日期 =(88×0.2422+22.83)-(88÷4)=

44-22=22，即 7 月 22 日。

例外：1922 年的计算结果加 1 日。

2. 节气特点

大暑是二十四节气中的第十二个节气，北半球在每年 7 月 22 ~ 24 日之间，南半球在每年 1 月 20 ~ 21 日之间，太阳位于黄经 120°。《月令七十二候集解》中说："暑，热也，就热之中分为大小，月初为小，月中为大，今则热气犹大也。"大暑节气正值"三伏天"里的"中伏"前后，是一年中最热的时候。

我国古代将大暑分为三候："一候腐草为萤；二候土润溽暑；三候大雨时行。"第一候是说陆生的萤火虫产卵于枯草上。大暑时，萤火虫卵化而出，所以古人认为萤火虫是腐草变成的。第二候是说天气开始变得闷热，土地也很潮湿。第三候是说时常有大范围的雷雨出现，大雨使暑湿减弱，天气开始向立秋过渡。

大暑节气是华南一年中日照最多、气温最高的时期，也是雨水最丰沛、雷暴最常见的时期。有谚语说："东闪无半滴，西闪走不及。"意思是在夏天午后，闪电如果出现在东方，雨不会下到这里；若闪电在西方，则雨势很快就会到来，想要躲避都来不及。大暑期间持续的日照、高温和充沛的降雨量都有利于农作物的生长，但是同小暑一样，此时也是旱、涝、风灾等自然灾害高发的季节。因此，要及时播种和收割，避免自然灾害造成损失。

3. 节气习俗

【饮食】大暑节气的民俗主要体现在吃的方面，这一时节的民间饮食习俗大致分为两种：一种是吃凉性食物消暑，另一种是吃热性食物去湿。

粤东南地区流传着一句谚语，"大暑吃仙草，活如神仙不会老"。仙草冻是用一种特殊的草做成的，它的茎和叶可以在晒干后做成香草果冻，有着神奇的降暑功效。而我国的台湾地区则有在大暑吃凤梨的习俗，因为这个时节的凤梨最好吃，而且有败火的作用。

大暑天气酷热，出汗多，脾胃活动相对较差。这时人会容易疲倦和食欲不振。而淮山药有补脾健胃和益气补肾的作用。多吃淮山药，可以促进消化，改善腰膝酸软，使人感到精力旺盛。

与此相反的是，有些地方的人们习惯在大暑时节吃热性食物。如福建人要吃荔枝和米糟来"过大暑"。荔枝是一种有营养的水果，含有葡萄糖和维生素。人们通常先在冷的井水中浸泡荔枝，然后再吃。据说大暑的荔枝和人参一样营养丰富。米糟是由发酵的糯米制成的。在大暑期间，人们用红糖来煮米糟，它能够大补元气。

"冬吃萝卜夏吃姜，不需医生开药方。"伏天喝姜汤，散寒祛暑、

开胃止泻。此外，湘中、湘北素有一种传统的进补方法，就是大暑吃童子鸡。人们相信大暑吃鸡可以祛除体内湿气。

【活动】在浙江台州，送大暑船是一种民间传统习俗，已经有数百年的历史了。大暑船上装满了各种各样的祭祀的动物，如猪、羊、鸡、鱼和虾。几十个渔民在游行队伍中轮流抬着这艘船。鼓声雷动，烟花在空中飞舞，街道两旁挤满了祈求祝福的人。一系列的仪式结束之后，大暑船被抬到了码头。然后，这艘船被拖出渔港，在海上烧毁。人们用这种仪式来祈求丰收和健康。

July

There are two solar terms in July—Slight Heat and Great Heat.

Slight Heat

1. Calculation Formula of Slight Heat

The calculation formula: (Y×D+C)-L

The interpretation of formula: Y=the latter two digits of the year number, D=0.2422, L= the number of leap year, 21st C=7.108, 20th C=7.928 ("C" stands for century.).

For example, the date of Slight Heat in 2088=(88×0.2422+7.108)-(88÷4)=28-22=6. That is to say, they day will fall on July 6th.

Exception: The date of Slight Heat in 1925 and 2016 would be the day after the calculating result.

2. Features of Slight Heat

Slight Heat (*Xiaoshu*) is the 11th term of the Twenty-four Solar Terms and falls on July 7th or July 8th, when the sun reaches the celestial longitude of 105°. *A Collective Interpretation of the Seventy-two Pentads* says, "*Xiaoshu* falls in the 6th lunar month. *Shu* means hotness. The heat can be divided into slight heat and great heat. At the beginning of this month, it is slight heat. While in the middle of this month, it is great heat. At this time, it belongs to the former." That is to say, *Xiaoshu* signifies the hottest period is coming but the extreme hot point has yet to arrive. In most parts of the country, the weather condition is in line with the above-mentioned circumstance. Crops throughout the country have entered a period of vigorous growth, and field management needs to be strengthened.

The average temperature in the southern China is about 33°C. In mid-July, the daily average temperature may begin to be higher than 30°C and the maximum daily temperature is higher than 35°C in the low-altitude valley of the southeast of China. In the northern part of the northwestern plateau, frost and snow can still be seen at this time, which is equivalent to the early spring season in southern China.

At the beginning of Slight Heat, the plum rainy period in the Jianghuai River ended, while the vast area north of Qinling Mountains and eastern Huaihe River begin to enter the southeast monsoon rainy season from the Pacific Ocean. Precipitation increases significantly, and rainfall is relatively concentrated. The southern part, southwestern part

and Qinghai-Tibet Plateau of China are also in the southwest monsoon rainy season from the Indian Ocean and South China Sea; in the middle and lower reaches of the Yangtze River, the weather is scorching with little rainfall under the control of the subtropical high pressure. There are also years when the cold air forces in the north before and after Slight Heat are still strong. In the middle and lower reaches of the Yangtze River, it is in neck-and-neck competition with the warm air in the south, which causes a frontal thunderstorm. Thunderstorms during the Slight Heat season are often the weather forecast of *daohuangmei*. It is an omen that the rainbands will remain in the middle and lower reaches of the Yangtze River for some time.

Storms, thunder and hail often happen around the Slight Heat in the most parts of southern China, though in some years there might be droughts. Thunderstorm is a kind of severe weather phenomenon, often accompanied by strong winds and heavy rain, and sometimes hail, which can easily cause disasters. In the eastern part of South China, due to the control of the subtropical high pressure after the Slight Heat, there are more consecutive sunny days and the summer drought begins. One of the prevailing farming activities during the Slight Heat is staying on top of flood control and drought relief.

3. Customs of Slight Heat

【Diet】There is an old custom of eating lotus root among common people. There are plenty of carbohydrates, vitamins and dietary fibers in

lotus roots. You will gain great benefits from having lotus roots with honey. It is appropriate for people to eat fresh lotus root in summer. The lotus root should be simmered over a low flame. After slicing it up, people need to add honey of appropriate amount. This dish can help soothe the nerves and sleep well.

There is a Chinese saying which goes that "the finless eel on the day of Slight Heat is as good as ginseng". Ricefield eels are habituated to live in the mud of ricefields, rivers, pools and lakes, and they are quite nutritious and delicious and they can function as the pills to cure some diseases. In addition, the content of iron is more than twice as high as that of carps and yellow croaker. Besides, it can also reduce the concentration of cholesterol in the blood and prevent cardiovascular diseases caused by arteriosclerosis. Therefore, it is pretty proper for people to eat ricefield eels on the day of Slight Heat.

There is a saying that it is hot in dog days. Slight Heat is the beginning of dog days. When the weather is hot, drink more soup or porridge, using such ingredients as lotus leaves, rhizoma smilacis glabrae, lentils, barley, grifolas, rhizoma alismatis, kapok, etc. It can be sweet or salty, which is very suitable for consumption on this solar term.

People in the north of China will eat mutton soup during the days of Slight Heat and Great Heat. It is said that one bowl of mutton soup will keep the doctors away. The first reason is nourishing the body. The second reason is that "羊" spelled in Chinese is a homophone for "阳". In ancient

times, people believed that eating mutton soup can help increase *yangqi* as much *yangqi* is consumed in summer.

【Event】During the Slight Heat period, many families hang their paintings and clothes out in the sun to prevent mildew. The old Chinese proverbs say, "on 6th in the 6th month, people hang out their clothes and dragons hang out robe", and "on 6th in the 6th month, every family hang out Red and Green". "Red and Green" refer to all kinds of colorful clothes. Because this day (basically on the eve before the Slight Heat) is the day with the highest temperature in the year, the longest sunshine time and the strongest sunlight radiation, the households choose this day to hang out clothes stored in the cabinets to receive sunlight.

There is a popular custom to taste new rice after the Slight Heat. The farmers ground the newly cut paddy into rice and make a meal to worship the great god of five cereals and ancestors, expressing their gratitude to nature and their ancestors. Finally, everyone can enjoy new wines.

Great Heat

1. Calculation Formula of Great Heat

The calculation formula: (Y×D+C)-L

The interpretation of formula: Y=the latter two digits of the year number, D=0.2422, L=the number of leap year, 21st C=22.83, 20th C=23.65("C" stands for century.).

For example, the calculating date of Great Heat in 2088 = (88×0.2422+

22.83)-(88÷4)=44-22=22, that is to say, the day will fall on July 22nd.

Exception: The date of Great Heat in 1922 would be the day after the calculating result.

2. Features of Great Heat

Great Heat (*Dashu*), which begins between July 22nd and July 24th in the Northern Hemisphere and January 20th and January 21st in the Southern Hemisphere each year, is the 12th term of the Twenty-four Solar Terms. On the day, the sun reaches the celestial longitude of 120°. *A Collective Interpretation of the Seventy-two Pentads* says, "*Shu* means hotness. The heat can be divided into slight heat and great heat. At the beginning of this month, it is slight heat. While in the middle of this month, it is great heat. At this time, it belongs to the latter." The Great Heat falls around the second of the three ten-day periods of the hot season, so it is the hottest time of the year.

In ancient China, the Great Heat was divided into three pentads: "The rotten grass is firefly, the soil is wet and heavy rain often appears." The first pentad is that terrestrial fireflies lay eggs on dead grass. On the day of Great Heat, fireflies spawn, so the ancients believed that fireflies were made of rotten grass; the second is that the weather begins to become sultry and the land is very humid; the third is that there are often large-scale thunderstorms. Heavy rain weakens the heat and the Start of Autumn is approaching.

The Great Heat is the period with the most sunshine and the highest

temperature in southern China. It is also the period with the most abundant rainfall and the most common thunderstorms. There is a saying goes, "There are no drops for the east lighting and you cannot able to go for the west lighting." It means that if lightning appears in the east in the summer afternoon, the rain will not come here; if lightning is in the west, the rain will soon come and it will be too late to escape. During the Great Heat period, the sunshine, high temperature and heavy rainfall are good for agricultural crops. But like the Slight Heat, many natural calamities such as floods, droughts and typhoons also happen during the Great Heat. Therefore, it's important to harvest and plant in time to avoid loss caused by natural disasters.

3. Customs of Great Heat

【Diet】There are two very contradictory customs on the Great Heat; one is to eat food with cold nature and the other is to eat food with warm nature.

In southeast of Guangdong Province, there is the tradition of eating "grass jelly". It is described in a popular saying, "Eating grass jelly in Great Heat will make you stay young like the immortals." The grass jelly is made with a special grass whose stems and leaves can be made into herb jelly after being dried in the sun. It has a special effect for relieving the summer heat. There is a custom in Taiwan which is eating pineapples in Great Heat. During this period, the pineapple tastes best and has the effect of relieving internal heat.

It is very hot on Great Heat. People will have a higher sweat output, so their spleen and stomach movements are relatively poor. At this time, people are prone to be tired and lose appetite. However, Chinese yam has the effect of invigorating spleen and stomach, reinforcing *qi* and tonifying kidney. Eating more Chinese yam can promote digestion and relieve the soreness of waist, making people feel energetic.

On the contrary, there is a tradition of eating food with warm nature on the day of Great Heat. In Fujian Province, people eat litchi and Mizao to celebrate the Great Heat. Litchi is a kind of nutritious fruit containing glucose and vitamins. People usually soak litchi in cold well water first and eat it. It is said that litchi during the Great Heat is as nourishing as ginseng. Mizao is made of fermented sticky rice. On the day of Great Heat, people cook them with brown sugar. It can reinforce the vital energy of the human body.

There is a folklore saying, "No need to bother the doctor for a prescription, if eating ginger in summer and radish in winter." People cook and drink a bowl of ginger tea to promote sweating in order to disperse wind-cold, improve stomach performance and digestion. In the middle and north of Hunan, people eat spring chicken. They believe that eating chicken on the day of Great Heat can drive away the dampness.

【Event】Sending the Great Heat ship is a folk tradition spanning hundreds of years in Taizhou, Zhejiang Province. The ship is filled with various animals for sacrifice such as pigs, sheep, chicken, fish and shrimps.

Decades of fishermen take turns carrying the ship as they march through the streets. Drums are played and fireworks are lit. Both sides of the street are filled with people praying for blessings. After a series of ceremonies, the ship is finally carried to the wharf. Then, the ship is pulled out of the fishing port and burned at sea. People carry out this ritual to pray for good harvests and health.

八月

八月有立秋、处暑两个节气。

立秋

1. 日期计算

计算公式：(Y×D+C)-L

公式解读：Y= 年数的后 2 位，D=0.2422，L= 闰年数，21 世纪 C=7.5，20 世纪 C=8.35。

举例说明：2088 年立秋日期 =(88×0.2422+7.5)-(88÷4)=28-22=6，即 8 月 6 日。

例外：2002 年的计算结果加 1 日。

2. 节气特点

立秋，是二十四节气中的第十三个节气，时为农历每年七月初一前后(公历 8 月 7 ~ 9 日之间)。立秋预示着炎热的夏天即将过去，

秋天即将来临，硕果金秋即将到来。据《月令七十二候集解》，立秋节气共有三候：“凉风至”“白露降”以及“寒蝉鸣”。到了秋天，梧桐树开始落叶，因此有“一叶知秋”的成语。从文字角度来看，“秋”字由禾与火字组成，是禾谷成熟的意思。秋季是天气由热转凉，再由凉转寒的过渡性季节，立秋是秋季的第一个节气。

3. 节气习俗

【饮食】在杭州，人们会在立秋当天吃桃子。在大家吃完桃子之后，将桃核一直保存到除夕，然后扔进炉子里烧成灰。人们相信在立秋的时候用这种方式可以全年预防瘟疫。

立秋时节可多食西兰花、海带、藕等食物。西兰花含有大量的维生素A、维生素C和胡萝卜素，对于提高皮肤抗损伤能力、保持肌肤弹性都起到了很好的作用。海带含有丰富的矿物质，可以调节血

液的酸碱度，防止皮肤过多分泌油脂。藕不论生熟，都具有很高的药用价值。藕粉既富于营养，又易于消化，有养血止血、调中开胃之功效，实为老幼体虚者理想的营养佳品。

【活动】在湖南、江西、安徽等地的山区，有“晒秋”习俗。因为平地少，农民便利用自家窗台、房顶架晒或挂晒农作物。“晒秋”的秋不仅指秋天，更寓意丰收和收获的果实。

在立秋第一天，人们通常会称体重，并和立夏时的体重作比较。如果体重下降了，那么秋天开始的时候，人们就会吃许多美味的食物来进补，特别是肉食，增加营养，补偿夏天的损失，这便叫做“贴秋膘”。在立秋这天流行吃各种各样的肉，如炖肉、烤肉、红烧肉等，“以肉贴膘”。

处暑

1. 日期计算

计算公式：(Y×D+C)-(Y÷L)

公式解读：Y= 年数的后 2 位，D=0.2422，L= 闰年数，21 世纪 C=23.13，20 世纪 C=23.95。

举例说明：2088 年处暑日期 =(88×0.2422+23.13)-(88÷4)=44-22=22，即 8 月 22 日。

2. 节气特点

处暑，是二十四节气之中的第十四个节气，时为公历 8 月 23 日前后，太阳到达黄经 150°。《月令七十二候集解》有言：“七月中，处，止也，暑气至此而止矣。”此时太阳正运行到了狮子座的轩辕十四星

近旁。夜观北斗七星，弯弯的斗柄还是指向"申"(西南方向)。"处暑"表示夏天的暑热日子正式画上句号，大部分地区即将进入秋天。同时，古人还将处暑划分为三候："一候鹰乃祭鸟；二候天地始肃；三候禾乃登。"

8月底到9月初的处暑节气，气温开始走低，原因之一是太阳的直射点继续南移，太阳辐射减弱；二是副热带高压向南撤退，蒙古冷高压开始跃跃欲试，小露锋芒。此后冷高压开始影响我国，形成的下沉、干燥的冷空气，先后宣告东北、华北、西北雨季终结，开始了一年之中最美好的天气——秋天。

"秋老虎"是指三伏出伏以后出现的短期高温天气。处暑时节，秋高气爽是常态，但是自古传到今的"秋老虎"也不是白叫的。夏季称雄的副热带高压，虽说大步南撤，但绝不肯轻易让出主导权。在它

控制的南方地区，刚刚感受一丝凉意的人们，往往在处暑尾声，再次感受高温天气，这就是名副其实的“秋老虎”。

长江中下游地区往往在“秋老虎”天气结束后，才会迎来秋高气爽的小阳春，不过要到10月以后了。在此期间，全国各地的暴雨总趋势是减弱的。但9月份仍是南海和西太平洋热带气旋活动较多的月份之一，热带风暴或台风带来的暴雨，对华南和东南沿海影响较大，降水强度一般呈现从沿海向内陆迅速减小的特点。急风暴雨带来洪水地质灾害仍需关注。

3. 节气习俗

【饮食】民间有处暑吃鸭子的习俗，因为老鸭味甘性凉，可以祛暑补虚、滋阴润燥。

“病从口入”，饮食与疾病很容易“挂钩”。处暑时节，像西瓜这类大寒的瓜果，则要少吃，宜多吃一些苹果、梨、葡萄之类的水果。

处暑之后，早晚温差变化开始明显，适宜进食清热安神的食物，如银耳、百合、莲子、芝麻、蜂蜜、乳制品、蔬菜、水果等。

【活动】处暑节气前后的民俗多与祭祖及迎秋有关。处暑前后民间会有庆祝中元节的民俗活动，俗称“七月半”。旧时民间从七月初一起，就有“开鬼门”的仪式，直到月底关鬼门止，都会举行普度布施活动。

处暑之后，秋意渐浓，正是人们畅游郊野迎秋赏景的好时节。处暑过，暑气止，就连天上的云彩也显得松散自如，而不像夏天大暑之时浓云成块。民间向来有“七月八月看巧云”之说，其间就有“出游迎秋”之意。

August

There are two solar terms in August—Start of Autumn and End of Heat.

Start of Autumn

1. Calculation Formula of Start of Autumn

The calculation formula: $(Y \times D + C) - L$

The interpretation of formula: Y =the latter two digits of the year number, D=0.2422, L=the number of leap year, 21st C=7.5, 20th C=8.35("C" stands for century.).

For example, the calculating date of the Start of Autumn in 2088 = $(88 \times 0.2422 + 7.5) - (88 \div 4) = 28 - 22 = 6$,that is to say, the day will fall on August 6th.

Exception: The date of the Start of Autumn in 2002 would be the day after the calculating result.

2. Features of Start of Autumn

Start of Autumn (*Liqiu*), which begins around the first day of the 7th lunar month (between August 7th and August 9th in the solar calendar), is the 13th term among the Twenty-four Solar Terms. The Start of Autumn reflects the end of summer and the beginning of autumn. The fruitful season is approaching. In autumn, the Chinese parasol trees begin to shed leaves, so there is an idiom that "a falling leaf heralds autumn". From a

textual point of view, the word "秋" consists of "禾" and "火", which shows that the cereals are mature. Autumn is the transitional season when the weather turns cool from the hot condition , and then turns cold from coolness. The Start of Autumn is the first solar term in autumn.

3. Customs of Start of Autumn

【Diet】In Hangzhou, people eat peaches on the day of Start of Autumn. The peach stones are kept until New Year's Eve and thrown into the stove, burned into ash. People believe that in this way, plagues can be prevented for the whole year.

It is good for people to eat broccoli, kelp and lotus root on the Start of Autumn. Broccoli contains a lot of vitamin A, vitamin C and carotene, which is beneficial for improving the skin's ability to resist damage and maintaining skin elasticity. Kelp is rich in minerals, which can regulate the pH value of the blood and prevent excessive secretion of skin sebum. Regardless of being cooked or raw, lotus root has a high medicinal value. The easily digestible lotus root powder is rich in nutrients with the effects of nourishing blood, promoting hemoslasis and whetting the appetite. It is truly an ideal nutritional product for the elderly and young.

【Event】In the mountainous areas of Hunan, Jiangxi and Anhui, there is a custom of "drying crops in autumn sunshine". Because of the lack of flat land, farmers use their own windows, roof shelves to dry crops in autumn sunshine. The autumn in "drying crops in autumn sunshine" refers not only to autumn itself, but also to the harvest and fruits reaped.

On the day of Start of Autumn, people usually weigh themselves and compare their weight to what it was at the Start of Summer. If one has lost weight, then at the beginning of autumn, he or she needs to flesh out by eating many different kinds of delicious food, especially meat. This can help increase nutrition and compensate for the weight loss in summer, that is, fleshing out in autumn. On the day of Start of Autumn, eating a variety of meat, such as stewed meat, roasted meat and braised pork in brown sauce, is a trend.

End of Heat

1. Calculation Formula of End of Heat

The calculation formula: (Y×D+C)-(Y÷L)

The interpretation of formula: Y =the latter two digits of the year number, D=0.2422, L=the number of leap year, 21st C=23.13, 20th C=23.95("C" stands for century.).

For example, the calculating date of End of Heat in 2088 = (88×0.2422+23.13)-(88÷4)=44-22=22, that is to say, the day will fall on August 22nd.

2. Features of End of Heat

End of Heat (*Chushu*), which falls on around August 23rd of the solar calendar, is the 14th solar term among the Twenty-four Solar Terms. At this time, the sun reaches the celestial longitude of 150°. *A Collective Interpretation of the Seventy-two Pentads* says, "*Chu* in Chinese stands for ending. In the middle of the 7th lunar month, the summer heat comes to an end." At the same time, the sun is moving to the point of Leo. The bent handle of the Big Dipper can still be seen pointing to the southwestern direction. End of Heat implies that most parts in China are getting rid of the hot summer and entering autumn. In ancient China, the End of Heat was divided into three pentads: "Eagles worship the birds, everything begins to withdraw and grains become ripe."

During the End of Heat season from the end of August to the early September, the temperature begins to fall. One of the reasons is that the solar radiation in China becomes weaker and weaker with the latitude of direct solar radiation; the other is that the subtropical high pressure moves southwards and the Mongolian cold high becomes stronger. The Mongolian cold high begins to influence China. Under its control, sinking and dry cold air is formed. It marks the ending of the rainy season in North China, Northeast China and Northwest China. Autumn, the best season in China begins.

"Autumn Tiger", known as an Indian Summer in English, signifies a period of hot weather during the traditionally cooler months. Usually, when autumn starts, the weather is supposed to get cooler. It does initially. Although the subtropical high has moved southwards, it still controls the southern part of China. The residual power of the summer will not just give up easily and it lingers into autumn. People in this area suffer from high temperature again even at the day of End of Heat. This is called the

spell of hot weather after the beginning of autumn.

It is necessary to mention that in the middle and lower reaches of the Yangtze River, cool weather won't come after the spell of hot weather until October. During this period, there are fewer rainstorms all over China. However, there are many tropical cyclones of South China Sea and the Western Pacific Ocean in September. The rainstorms brought about by tropical storms or typhoons have a great impact on the South and Southeast China coasts and the rainfall intensity generally shows the characteristics of rapid reduction from the coast to the inland. Floods and geological disasters caused by torrential winds and rain still need attention.

3. Customs of End of Heat

【Diet】Duck has a sweet flavor and according to Chinese traditional medicine it has a "cool" nature. A folk tradition is to eat duck during the End of Heat period.

There is an old idiom, "Out of the mouth comes evil." It can be seen that diet and disease are easily connected. People should eat more seasonal fruit like apples, pears and grapes in the End of Heat season, but less watermelons which are with cold nature.

After the End of Heat, the temperature varies from the morning to the evening. During this season, it's better to take more food which can clear heat fire, ease thirst and soothe the nerves, such as tremella, lilies, lotus seeds, sesame, honey, dairy product, vegetables, fruits, etc.

【Event】The folk customs before and after End of Heat are mostly related to ancestor worshipping and autumn greeting. An event called Zhongyuan Festival falls on the 15th of the 7th lunar month. In ancient times, a ceremony called "opening the ghost door" began on the 1st of the 7th lunar month and lasts to the end of the 7th lunar month. During that period, people conduct activities to help ghosts reincarnate.

After the coming of End of Heat, people like to go on an excursion and enjoy the early autumn scenery and welcome the coming of autumn. When the End of Heat comes, summer heat is gone. Clouds in the sky scatter around forming different shapes. There is a saying that goes, "Enjoying the clouds of various forms in the 7th and 8th lunar month." After the End of Heat, people can enjoy the scenery as it changes gradually from summer to autumn.

九月

九月有白露、秋分两个节气。

白露

1. 日期计算

计算公式：(Y×D+C)-L

公式解读：Y= 年数的后 2 位，D=0.2422，C= 闰年数，21 世纪 C=7.646，20 世纪 C=8.44。

举例说明：2013 年白露日期 =(13×0.2422+7.646)-(13÷4)=10.7946-3.25=7.5，9 月 7 日是白露。

例外：1927 年的计算结果加 1 日。

2. 节气特点

白露是二十四节气中的第十五个节气，时为在公历每年 9 月 7 日到 9 日，太阳到达黄经 165° 时。《月令七十二候集解》中说："八月节……阴气渐重，露凝而白也。"白露是九月的第一个节气。露是由于温度降低，夜晚水汽在草地或树上凝结而成的水珠。所以，白露实际上是表示天气已经转凉。这时，人们就会明显地感觉到炎热的夏天已过，而凉爽的秋天已经到来了。

俗语云："处暑十八盆，白露勿露身。"这两句话的意思是说，处暑仍热，每天需用一盆水洗澡，过了十八天，到了白露，就不要赤膊裸体了，以免着凉。

3. 节气习俗

【饮食】民间有“春茶苦，夏茶涩，要喝茶，秋白露”的说法。白露时节的茶树经过夏季的酷热，此时正是它生长的最佳时期。白露茶既不像春茶那样鲜嫩、不经泡，也不像夏茶那样干涩味苦，而是有一种独特的甘醇清香，尤受老茶客喜爱。

白露时节还可以吃梨、苹果、黄瓜、柑橘、胡萝卜、石榴、龙眼等。

梨水分丰富，同时有生津止渴、止咳化痰、清热降火、养血生肌、润肺去燥等功效。苹果富含果胶、纤维素、维生素C等。黄瓜生吃可以充分补充皮肤所含的水分，黄瓜片敷脸可以滋养皮肤。柑橘性温、味甘，有开胃理气、止咳润肺的功效。胡萝卜味甘、性平，具有清热解毒的功效，适宜于皮肤干燥粗糙，或患毛发苔藓、黑头粉刺、角化型湿疹者食用。石榴性温，具有生津液、止烦渴的作用。

白露时，龙眼完全成熟，甜度最高，口感最好，福州人这个时节喜欢吃龙眼。传说“白露过后吃龙眼，一颗龙眼顶只鸡”。虽然夸张，但龙眼肉甘温滋补，入心脾两经，功善补益心脾，而且甜美可口，实为补心健脾之佳品，更是白露进补的好选择。

在南方，历来有白露时节酿酒的习俗。白露米酒是用高粱、糯米之类的五谷酿成，略带甜味。“程酒”是白露米酒中的精品，因取程江之水酿制而得名。过去家家酿酒，酿好的白露米酒是待客的必备品之一。

【活动】白露时节是太湖人祭禹王的日子。禹王是传说中的治水英雄大禹，太湖畔的渔民称他为“水路菩萨”。每年正月初八、清明、七月初七和白露时节，这里将举行祭禹王的香会，其中又以清明、白露春秋两祭的规模为最大，历时一周。

在祭禹王的同时，还祭土地神、花神、门神、宅神、姜太公等。祭祀活动期间，《打渔杀家》是必演的一台戏，它寄托了人们对美好生活的祈盼和向往。

秋分

1. 日期计算

计算公式：(Y×D+C)-L

公式解读：Y= 年数的后 2 位，D=0.2422，L= 闰年数，21 世纪 C=23.042，20 世纪 C=23.822。

举例说明：2088 年秋分日期 =(88×0.2422+23.042)-(88÷4)= 44-22=22，即 9 月 22 日。

例外：1942 年的计算结果加 1 日。

2. 节气特点

秋分，农历二十四节气中的第十六个节气，时为每年的9月22或23日。南方由这一节气起才始入秋。《春秋繁露》记载："秋分者，阴阳相半也，故昼夜均而寒暑平。""秋分"的含义有两层。一是按我国古代以立春、立夏、立秋、立冬为四季开始划分四季，秋分日居于秋季90天之中，平分了秋季。二是昼夜均分。秋分日，太阳到达黄经180°，直射地球赤道，这一天24小时昼夜均分。秋分之后，太阳直射点南移，北半球各地开始昼短夜长，即一天之内白昼开始短于黑夜。《月令七十二候集解》将秋分分为三候："一候雷始收声；二候蛰虫坯户；三候水始涸。"

按气候学上的标准，秋分时节，我国长江流域及其以北的广大地区，日平均气温都降到了22℃以下。此时，全国大部分地区进入凉爽的秋季，气温持续下降，降雨量减少，空气干燥，东北地区见霜已不足为奇。

3. 节气习俗

【饮食】秋分时节，寒凉渐重，多出现凉燥，饮食应以清润、温润的食物为主，润养成分最多的粥汤，如甘蔗粥、百合粥、栗子粥、胡萝卜粥、冰糖银耳汤等，均有清热润燥、益气生津的功效，且口感味道极好，为秋分当令美食。

【活动】秋分曾是传统的"祭月节"。现在的中秋节则是由传统的"祭月节"而来。据考证，最初祭月节是定在秋分这一天。不过由于这一天不一定都有圆月，而祭月无月则是大煞风景的。所以，祭月节后来就由秋分调至中秋。

秋分时节，农村便有挨家送秋牛图的。这些图是在红纸或黄纸

上印上全年农历节气，还要印上农夫耕田图样，名曰“秋牛图”。送图者都是些民间善言唱者，主要说些秋耕和吉祥不违农时的话，每到一家，说到主人乐而给钱为止。这一活动俗称“说秋”，说秋人便叫“秋官”。

September

There are two solar terms in September—White Dew and the Autumn Equinox.

White Dew

1. Calculation Formula of White Dew

The calculation formula: (Y×D+C)-L

The interpretation of formula: Y= the latter two digits of the year number, D=0.2422, L=the number of leap year, 21st C=7.646, 20th C=8.44("C" stands for century.).

For example, the date of White Dew in 2013=(13×0.2422+7.646)-(13÷4)=10.7946-3.25=7.5. That is to say, the day would fall on September 7th.

Exception: The date of White Dew in 1927 would be the day after the calculating result.

2. Features of White Dew

White Dew (*Bailu*) is the 15th term among the Twenty-four Solar

Terms and falls between September 7th and September 9th every year, when the sun reaches the celestial longitude of 165°. *A Collective Interpretation of the Seventy-two Pentads* states, "In the eighth month of lunar calendar... *yinqi* gradually increases and White Dew appears." White dew is the first solar term in September. White Dew indicates the real beginning of cool autumn. The temperature declines gradually and the vapors in the air often condense into white dew on the grass and trees at night. This is the time when most regions of South China undergo the continuing dropping of temperature. People clearly feel that the hot summer has passed and the cool autumn has arrived.

The saying goes: "Eighteen basins are needed on End of Heat; during White Dew, the body should not be exposed." The meaning of these two sentences is that it is still hot on End of Heat and it is necessary to bathe with a basin of water every day. After 18 days when White Dew arrives, people cannot expose their body so as not to catch cold.

3. Customs of White Dew

【Diet】There is a folk saying: "Spring tea is bitter and summer tea is astringent; if people wants to drink tea, choose White Dew tea." Tea during White Dew has gone through the hot summer and is in its best state of growth. White Dew tea is different from spring tea which is usually too tender and cannot resist hot water for long. It is also different from summer tea, which is dry and has a bitter flavor. White Dew tea tastes sweet with its fragrance.

People can eat pear, apple, cucumber, citrus fruit, carrot, pomegranate, longan and so on in the season of White Dew.

The pear, which is rich in water content, has the functions of producing saliva and quenching thirst, relieving cough and phlegm, reducing heat, nourishing blood and promoting tissue regeneration as well as moistening lungs. Apple is rich in pectin, cellulose, vitamin C and so on. Eating raw cucumber can fully supplement the moisture contained in the skin. In addition, putting cucumber slices on the face can nourish the skin. The sweet citrus fruit is warm in nature with the effect of whetting the appetite, regulating the flow of vital energy, relieving cough and moistening lungs. The sweet carrot, which is neutral in nature, has the effect of clearing heat and removing toxicity. It is suitable for people with dry and rough skin or patients suffering from lichen pilaris, comedones and keratinized eczema to eat. Pomegranate is warm in nature and has the effect of producing saliva and quenching thirst.

Longan around White Dew is big, sweet and tastes great. It is a tradition in Fuzhou, Fujian province that eating longan on the first day of White Dew, which can help nourish the human body. It is said that one longan is as nutritious as chicken. Although this sounds exaggerated, with a sweet taste, longan does reinforce the spleen, nourish the blood, calm the nerves and improve one's looks, which is a good choice for the White Dew.

It is traditional to make White Dew Wine during this season in southern China. The wine is made of cereals such as polished glutinous rice and kaoliang (a specific type of sorghum) and tastes a little sweet. Cheng Wine is the best among all of them, which is named after its source—water from Cheng River. In the past, country people in those regions all made wine and served it to guests when entertaining.

【Event】People from Taihu usually worship Yu the Great during White Dew season. Yu the Great is a legendary hero of flood control, and the fishermen on the banks of Taihu Lake call him the "Bodhisattva of Water". The incense-offering fair in commemoration of Yu the Great will be held here four times every year, respectively falling on the eighth day of the first lunar month, Qingming Festival, the seventh day of the seventh lunar month and White Dew season. Among which, the fairs on Qingming Festival and White Dew season are the largest ones, which will last for a week.

While worshipping Yu the Great, people also offer sacrifices to the God of land, the God of flower, the God of door, the God of house, Jiang Taigong and so on. During the sacrificial activities, *A Fisherman's Revenge* is a play that must be performed, which symbolizes people's hope and longing for a good life.

Autumn Equinox

1. Calculation Formula of Autumn Equinox

The calculation formula: (Y×D+C)-L

The interpretation of formula: Y=the latter two digits of the year

number, D=0.2422, L=the number of leap year, 21st C=23.042, 20th C=23.822("C" stands for century.).

For example, the date of Autumn Equinox in 2088=(88×0.2422+23.042)-(88÷4)=44-22=22. That is to say, the day will fall on September 22nd.

Exception: The date of the Autumn Equinox in 1942 would be the day after the calculating result.

2. Features of Autumn Equinox

Autumn Equinox (*Qiufen*), the 16th term among the Twenty-four Solar Terms of the lunar calendar, begins on September 22nd or September 23rd each year. Since the Autumn Equinox, most of the areas in China have entered the cool autumn. As it is said in the ancient book, *The Detailed Records of the Spring and Autumn Period*, "It is on the Autumn Equinox day that the yin and yang are in a balance of power. Thus the day and night are of equal length, and so are the cold and hot weather." The word *Qiufen* has two meanings. One is that the Autumn Equinox lies at the midpoint of autumn, dividing autumn into two equal parts. Here the autumn refers to the 90 days from the Start of Autumn to the Start of Winter. (In ancient times, the four seasons begin with the Start of Spring, the Start of Summer, the Start of Autumn and the Start of Winter respectively.) The other is that it equally divides the day and night. On this day, the sun reaches the celestial longitude of 180° and shines directly at the equator of the earth. After that day, the location of direct sunlight moves to the south, making days shorter and nights longer in the northern hemisphere. According to *A Collective Interpretation of the Seventy-two Pentads*, the Autumn Equinox is divided into three pentads: “Thunder begins to decrease, insects make homes and water begins to solidify.”

Climatologically, the average daily temperature in the Yangtze River basin and its vast area north of China falls below 22°C during the Autumn Equinox season. At this time, the cool autumn period begins in most areas of China and the temperature steadily drops. The amount of rainfall decreases and it’s getting drier and drier in the air. In some part of the northeast China, it’s not strange to see frost at this time.

3. Customs of Autumn Equinox

【Diet】On the Autumn Equinox season, it is getting colder. People are more likely to suffer from cool-dryness syndrome. So their diet should be fresh, moist food. The most nutritious porridge and soup, such as sugar cane porridge, lily porridge, chestnut porridge, carrot porridge, snow fungus soup added with rock candy, all have the effect of relieving internal heat and dryness, replenishing *qi* and promote the secretion of body fluid. Furthermore, the taste is also excellent in this season.

【Event】Autumn Equinox is the time to offer sacrifices to the Moon. According to historical research, the original moon ritual is held on the day of Autumn Equinox. However, since it doesn’t happens on the same day each year and sometimes the moon on this day is not the roundest. A ceremony without full moon is not a perfect one. Later, the ceremony date

was changed from Autumn Equinox to the Mid-Autumn Festival.

On the day of Autumn Equinox, red or yellow pictures of cattle farming in the field were considered auspicious gifts by farmers. Some villagers were good at speaking and singing work as *qiuguan* (autumn official) for *shuoqiu* (speaking about autumn) and gave away paintings to every household in the village. They were expected to speak auspicious words highlighting the farming season, while the families gave away money when they believed the words would brings luck and joy.

十月

十月有寒露、霜降两个节气。

寒露

1. 日期计算

计算公式：(Y×D+C)-(L÷4)

公式解读：Y= 年数的后 2 位，D=0.2422，L= 闰年数 21 世纪 C=8.318，20 世纪 C=9.098。

举例说明：2088 年寒露日期 =(88×0.2422+8.318)-(88÷4)= 29-22=7，即 10 月 7 日。

2. 节气特点

寒露是二十四节气中的第十七个节气，时间为每年的 10 月 7 ~

9 日，属于秋季。太阳的直射点在南半球继续南移，北半球阳光照射的角度开始明显倾斜，地面所接收的太阳热量比夏季显著减少，冷空气的势力范围所造成的影响，有时可以扩展到华南。据《月令七十二候集解》："九月节，露气寒冷，将凝结也。"意思是寒露的气温比白露时更低，地面的露水更冷，快要凝结成霜了。寒露时节，南岭及以北的广大地区均已进入秋季，东北进入深秋，西北地区已进入或即将进入冬季。

我国古代将寒露分为三候："一候鸿雁来宾；二候雀入大水为蛤；三候菊有黄华。"此节气时，鸿雁排成一字或人字形的队列大举南迁；深秋天寒，雀鸟都不见了，古人看到海边突然出现很多蛤蜊，并且贝壳的条纹及颜色与雀鸟很相似，所以便以为是雀鸟变成的；第三候的"菊始黄华"是说在此时菊花已普遍开放。寒露之后，露水增多，气温

更低。此时我国有些地区会出现霜冻，北方已呈深秋景象，白云红叶，偶见早霜，南方也秋意渐浓，蝉噤荷残。

3. 节气习俗

【饮食】寒露时节，应根据个人的具体情况，适当进补，既可补脾胃，又能养肺润肠。在饮食上还应少吃辛辣刺激、香燥、熏烤等类食物，宜多吃些核桃、银耳、萝卜等有滋阴润燥、益胃生津作用的食物，还应多吃雪梨、苹果等水果。早餐应吃温食，最好喝热药粥，比如甘蔗粥、沙参粥等，因为粳米、糯米均有极好的健脾胃、补中气的作用。中老年人和慢性病患者应多吃些红枣、莲子、山药、鸭、鱼、肉等食物。

当秋之时，要减辛味食物以平肺气，增酸味食品以养肝气，防肺气太过乘肝，使肝气郁结。因此，在饮食上，要尽可能少食葱、姜、蒜、韭等辛味之品，多吃一点酸味果蔬。酸味可防止能量损失，并具有收缩和收敛的作用。在生鱼片上滴些柠檬汁后，生鱼片会变得有质感，就是酸味收敛作用的体现。酸味收敛补肺，故秋季宜多吃酸的水果，比如葡萄、橘子、山楂等。

俗话说："寒露收山楂，霜降刨地瓜。寒露柿红皮，摘下去赶集。"柿子营养价值很高，含有丰富的蔗糖、葡萄糖、果糖、蛋白质、胡萝卜素、维生素C等营养物质。其所含维生素和糖分比一般水果高1～2倍，自一般一天最好不要食用超过一个。

寒露节，菊花盛开，为除秋燥，许多地方有饮"菊花酒"的习俗。寒露与重阳节接近，这一习俗与登高一起，渐渐移至重阳节。菊花酒，古称"长寿酒"，是由菊花加糯米、酒曲酿制而成，其味清凉甜美，有延缓衰老的功效。

由于天气渐冷，树木花草凋零在即，故人们谓此为"辞青"。

九九登高时，还要吃花糕，因“高”与“糕”谐音，故应节糕点谓之“重阳花糕”，寓意“步步高升”。

【活动】寒露时节是钓鱼的好时节，因为，进入白露以后，天气已逐渐凉爽，水温也下降到鱼类喜爱的温度。此时，饱受了盛夏苦日的鱼儿又开始活跃起来，四处游弋觅食。尤其在寒露、霜降节气的晚秋，鱼儿为填肚越冬要贮存食料就显得更加馋嘴贪吃，易上钩、易钓获。

寒露到来的农历九月又称菊月，是菊的月份。越是霜寒露重，菊花开得越艳丽。寒露与霜降节气，是观赏菊花的最佳时节。

霜降

1. 日期计算

计算公式：(Y×D+C)-L

公式解读：Y= 年数的后 2 位，D=0.2422，L= 闰年数，21 世纪 C=23.438，20 世纪 C=24.218。

举例说明：2088 年霜降日期 =(88×0.2422+23.438)-(88÷4)=44-22=22，10 月 22 日霜降。

例外：2089 年的计算结果加 1 日。

2. 节气特点

霜降是二十四节气中的第十八个节气，时为每年阳历 10 月 23 日前后，太阳到达黄经 210° 时。霜降是秋季的最后一个节气，是秋季到冬季的过渡节气。秋天夜晚地面上散热很多，温度骤降到 0° 以下，空气中的水蒸气在地面或植物上直接凝结形成细微的冰针，有的成为六角形的霜花，色白且结构疏松。《月令七十二候集解》关于霜降有言：“九月中，气肃而凝，露结为霜矣。”“霜降”表示天气逐渐变冷，

露水凝结成霜。我国古代将霜降分为三候:“一候豺乃祭兽;二候草木黄落;三候蛰虫咸俯。”豺这类动物从霜降开始要为过冬储备食物,树叶枯黄掉落,蛰虫也全在洞中不动不食,垂下头来进入冬眠状态。霜降节气到来天气越来越冷,吃完了凌霜的柿子,就该进入冬天了。

3. 节气习俗

【饮食】霜降节气的到来意味着冬季将要到来,饮食上首先推荐白萝卜。中医认为,白萝卜性凉、味甘,具有清热生津、开胃健脾、化痰之功效。白萝卜中含有丰富的维生素 C 和微量元素,可帮助机体增强免疫力,促进肠胃蠕动,以增加食欲。尤其在冬天,人们常常会出现痰多、肺部不适等症状,适量地吃一些白萝卜,可起到辅助治疗的作用。

白果，又叫银杏，是营养丰富的高级滋补品，霜降时节经常食用白果，可以滋阴养颜抗衰老，扩张微血管，促进血液循环，使人肌肤红润、精神焕发，另外它还具有敛肺定喘、燥湿止带等功效。不过白果虽然美味，却不能多吃，因白果含有少量氰化物，不可长期、大量生食，以免中毒。

鲜橄榄入口虽苦，但是经过咀嚼后，越发甘甜，是秋冬防燥润喉的佳品。橄榄在医学上用为清肺利咽药。霜降天气干燥，嚼食鲜橄榄可以润喉、清热、止渴、生津。

栗子具有养胃健脾、补肾强筋、活血止血、止咳化痰的功效，是霜降节气的进补佳品。它对肾虚有良好的疗效，故又称为“肾之果”。熟食板栗能和胃健脾，缓解脾虚。将板栗仁蒸熟、磨粉，制成糕饼，适用于身体瘦弱的儿童，以增加食欲。

霜降要多吃牛肉。我国国民消费的肉类食品之中，牛肉蛋白质含量高，而脂肪含量低，味道鲜美，受人喜爱，享有“肉中骄子”的美称。中医认为牛肉有补中益气、滋养脾胃、强健筋骨、化痰息风、止渴止涎的功效，适宜于中气不足、气短体虚、筋骨酸软、久病贫血、面黄体瘦、头晕目眩的病人食用。适宜用量每餐约80克，过量食用可能会增加结肠癌和前列腺癌的患病几率。老年人、儿童和消化力弱的人不宜多吃牛肉。

步入霜降后因为干燥的天气，很多人会出现口干、咳嗽的症状，所以此时应该吃些具有润肺止咳功效的食物，百合就是不二之选。百合味道清香，可以用来炒菜、熬粥、煲汤等，并且可以起到润肺止咳的功效，是秋季滋补的佳品。吃百合的方法很多，可以当菜肴吃，如荠菜炒百合，先将荠菜切成末，与百合瓣共炒，称得上是美味佳肴，特

别适合于肺病患者食用。

【活动】古有“霜打菊花开”之说，所以登高山，赏菊花，也就成为了霜降这一节令的雅事。南朝吴均的《续齐谐记》形容菊花，“霜降之时，唯此草盛茂”，因此菊被古人视为“候时之草”，成为生命力的象征。霜降时节正是秋菊盛开的时候，我国很多地方在这时要举行菊花会，赏菊饮酒，以示对菊花的崇敬和爱戴。古人眼里，菊花有着不寻常的文化意义，被称为“延寿客”“不老草”。

霜降时节人们有登高远眺的习俗。登高既可使身体得到锻炼，同时登至高处，天高云淡，枫叶尽染，极目远眺，赏心悦目。但也要有所讲究。登高的时间要避开气温较低的早晨和傍晚。登高时，要沉着，速度要慢，以防腰腿扭伤；下山不要走得太快，以免膝关节受伤或肌肉拉伤。登高过程中，应通过增减衣服来适应温度的变化；休息时，不要坐在潮湿的地上和风口处；出汗时可稍松衣扣，不要脱衣摘帽，以防伤风受寒。对于老年人来说，应带根手杖，这样既省体力，又有利于安全。在爬山时要注意力集中，并注意脚下石头是否活动，以免踏空。在陡坡行走时，最好采取“之”字形路线攀登，这样可缓解坡度。

October

There are two solar terms in October—Cold Dew and Frost's Descent.

Cold Dew

1. Calculation Formula of Cold Dew

The calculation formula: $(Y \times D + C) - (L \div 4)$

The interpretation of formula: Y=the latter two digits of the year number, D=0.2422, L=the number of leap year, 21st C=8.318, 20th C=9.098("C" stands for century.).

For example, the date of the Cold Dew in 2088=(88×0.2422+8.318)-(88÷4)=29-22=7. That is to say, the day will fall on October 7th.

2. Features of Cold Dew

Cold Dew (*Hanlu*), the 17th term among the Twenty-four Solar Terms, begins on October 7th to 9th. It belongs to autumn. The subsolar point continues to move southwards in the southern hemisphere. The angle of sunlight in the northern hemisphere begins to tilt obviously. The solar heat received by the ground significantly reduces than in summer. The impact of cold air sphere can sometimes extends to southern China. *A Collective Interpretation of the Seventy-two Pentads* records, "Cold Dew in the 9th lunar month is colder than White Dew and the dews are becoming frost." During this period, most of the areas in the north of Nanling have entered autumn, the northeast of China has entered late autumn and the northwest of China has entered or will soon enter winter.

In ancient times, the Cold Dew period has been divided into three pentads: the first is that "wild geese migrate southward" in one line or lambdoid queue; the second is that "sparrows enter the ocean and become clams" when sparrows and clams were regarded as the same animal because of their similar stripes and colors; the third is that "chrysanthemums are in full bloom". After the Cold Dew, the dew is more and it is getting colder

than before. In some areas, the dew is about to freeze into frost. North China has taken on a look of late autumn with white clouds, red leaves and early frost. In southern China, cicadas keep silent and lotuses become withered.

3. Customs of Cold Dew

【Diet】In the Cold Dew season, nourishing food should be eaten according to the specific circumstances of the individual, which cannot only invigorate the spleen and stomach, but also nourish the lungs and intestinal tracts. What's more, people should eat less spicy, irritating, dried, smoked and other kinds of food. The food that can enrich yin and relieving dryness, nourish the stomach and produce such as saliva walnuts, tremella, radish can be good choices. Moreover, snow peas, apples and other fruits are also fit to eat. When it comes to breakfast, people should eat warm food. They had better have hot medicated porridge such as sugar cane porridge and straight ladybell porridge, because rice and glutinous rice have excellent effect of invigorating spleen, stomach and nourishing vitality. Elderly people and chronic patients should eat more red dates, lotus seeds, Chinese yams, ducks, fish, meat and other foods.

When autumn descends, eat less pungent food to balance lung *qi* and more acidic food to increase liver *qi* for fear that the higher content of lung *qi* over liver *qi* might lead to stagnation of liver *qi*. Therefore, people should eat less pungent food like onion, ginger, garlic and leek. They might as well eat more sour fruits and vegetables. Tart flavour can prevent energy

loss, so it has the effect of contraction and convergence. After dripping some lemon juice on the sashimi, sashimi will become textured, which is the embodiment of convergence. Furthermore, this antioxidants in fruits like grapes, oranges and hawthorns are thought to boost heart health by strengthening blood vessels and stimulating blood flow.

There is a saying goes, "Hawthorns are harvested on Cold Dew and sweet potatoes are reaped on Frost's Descent. The persimmons turn mature on Cold Dew. People had better pick these in time to sell in the market." Persimmons have high nutritional value and are rich in nutrients such as sucrose, glucose, fructose, protein, carotene and vitamin C. The content of vitamin and sugar is about 1 to 2 times higher than that other fruits contained in average, but it is best to eat no more than one persimmon a day.

Chrysanthemum is a symbol of Cold Dew. To prevent autumn dryness, many regions in China have the custom of drinking chrysanthemum wine which was called "longevity wine" in ancient times. This is a tradition of the Double Ninth Festival (on the ninth day of the ninth month in the Chinese calendar) besides climbing mountains, which often falls around Cold Dew. According to ancient records, drinking the wine made with chrysanthemums, glutinous rice and distiller's yeast grants people long lasting youth.

Because the weather is getting cold, trees and flowers wither. Therefore, people call this "saying goodbye to the green plants". On the Double

Ninth Festival, people climb mountains and eat flower cakes. Since "高" is a homophone for "糕", the cakes eaten during this day is addressed as "Chongyang flower cake", which means "promoting to a higher position".

【Event】The Cold Dew season is a good time for fishing. The weather gradually becomes cool after the arrival of White Dew and the water temperature has also dropped to the extent that fishes enjoy it. At this time, the fish that suffered from the bitter summer begins to be active again, swimming around for food. Especially in the late autumn of Cold Dew and Frost's Descent, the fishes seem more greedy for getting the over-wintering food, so it is easy to catch them.

Cold Dew begins in the ninth month of the lunar calendar. This month is also known as the chrysanthemum month. Chrysanthemum can bloom in greater beauty with more frost and dew. Cold Dew and Frost's Descent are the best time to appreciate chrysanthemum.

Frost's Descent

1. Calculation Formula of Frost's Descent

The calculation formula: (Y×D+C)-L

The interpretation of formula: Y=the latter two digits of the year number, D=0.2422, L=the number of leap year, 21st C=23.438, 20th C=24.218("C" stands for century.).

For example, the date of Frost's Descent in 2088=(88×0.2422+23.438)-(88÷4)=44-22=22. That is to say, the day will fall on October 22nd.

Exception: The date of Frost's Descent in 2089 will be the day after the calculating result.

2. Features of Frost's Descent

Frost's Descent (*Shuangjiang*), the 18th term among the Twenty-four Solar Terms, begins around October 23rd of the solar calendar. On this day, the sun reaches the celestial longitude of 210°. Frost's Descent is the last solar term in autumn, indicating the transition from autumn to winter. During the autumnal nights, a lot of heat on the ground is released and the temperature suddenly drops below 0°. The water vapor in the air condenses directly on the ground or on plants to form subtle ice needles. Some become hexagonal frost flowers, white in color and loose in structure. *A Collective Interpretation of the Seventy-two Pentads* mentions, "In the middle of the 9th lunar month, the air is condensed and the dew is frosted." "Frost's Descent" means that the weather is getting cold gradually and the dew turns slowly to be white frost. The Frost's Descent Festival is divided into three pentads: "Jackals reserve their preys, the plants become yellow and shed leaves and all insects go dormant." During the period, beasts reserve their prey before eating; the yellow leaves on the earth fall; the insects also stay in the hole without moving and lower their heads to enter hibernation. When Frost's Descent arrives, the weather gets colder and colder. After eating the persimmons that withstand frost, it is time to enter winter.

3. Customs of Frost's Descent

【Diet】The arrival of Frost's Descent means that winter will come.

White radish is highly recommended if people want to keep in good health at this time. According to traditional Chinese medicine, the sweet white radish is cold in nature with the effect of clearing heat and producing saliva, whetting the stomach, nourishing spleen and reducing phlegm. The white radish is rich in vitamin C and trace elements, which can help strengthen immunity, promote gastrointestinal peristalsis and work up an appetite. In winter, people often have symptoms such as excessive phlegm and lung discomfort. They should eat some white radishes to get assisted treatment.

Ginkgo is nutritious high-level tonic. Eating ginkgo frequently can nourish the skin to resist aging, expand capillaries, promote blood circulation, make human skin ruddy and refresh the mind. In addition, it also has the effects of astringing lung for relieving cough, eliminating dampness and checking vaginal discharge. However, although the ginkgo is delicious, it cannot be eaten too much. Because the ginkgo contains a small amount of cyanide, it cannot be eaten raw in a long term and for a large amount so as to avoid poisoning.

The olive tastes bitter, but it turns sweet after careful chewing. It is a good product for preventing dry throat in autumn and winter. Olives are medically used as medicine for clearing away the lung heat and relieving sore throat. The weather is dry in autumn. Chewing fresh olives can moisten the throat, clear heat, quench thirst and produce saliva.

Eating chestnuts during Frost's Descent is beneficial for one's health. Chestnuts have a warm nature and sweet flavor, and are good for

nourishing the spleen and stomach, invigorating the circulation of blood, relieving coughs and reducing sputum. Therefore, chestnut is also called "the fruit of the kidney". Cooked chestnut can nourish the stomach, spleen and relieve spleen deficiency. In addition, people can steam the chestnut kernels and grind these into powder, then make into cakes. These cakes are suitable for weak children to whet their appetite.

Beef, one of the meat consumed by the Chinese, should be eaten more on Frost's Descent. The high protein and low fat make it delicious and popular. It enjoys the reputation of "arrogant meat". Traditional Chinese medicine believes that beef has the effects of replenishing *qi*, nourishing the spleen and stomach, strengthening the muscles and bones, reducing phlegm and quenching thirst. It is suitable for patients suffering from deficiency of *qi* in the middle energizer, shortness of breath, weak bones, chronic anemia, emaciation with sallow complexion and dizziness. The appropriate ingestible amount is about 80 grams per meal. Excessive consumption may increase the risk of colon cancer and prostate cancer. Elderly people, children and people with weak digestion should not eat too much beef.

After entering the Frost's Descent, because of the dry weather, many people will have symptoms of dry mouth and cough. Therefore, we should eat some food that has the effects of moistening lungs and relieving cough. Lily is the best choice, and the taste is fragrant. It can be used for fried vegetable, porridge, soup and so on. And it can play the effects of moistening lungs and relieving cough. It is a good product for nourishing

the body. There are many ways to eat lilies. You can eat them as dishes, such as fried lilies. First cut shepherd's purse into pieces, and fry them with lilies. It is a delicious dish and is particularly suitable for consumption by patients with lung diseases.

【Event】There is an ancient saying: "Chrysanthemums bloom on Frost's Descent." Therefore, climbing mountains and admiring chrysanthemums have become a trend for this festival. In the Southern Dynasty, Wu Jun mentioned it in his book *The Witty and Mythical Stories Recorded from Qi Dynasty*, "When Frost's Descent falls, only this plant is in full bloom." Therefore, the chrysanthemum was regarded by the ancients as "the grass of time" and became a symbol of vitality. The Frost's Descent season is the time when autumn chrysanthemum is in full bloom. People in many places of China will hold chrysanthemum parties at this time and drink chrysanthemum tea to show respect and love for chrysanthemum. In the eyes of the ancients, chrysanthemum had unusual cultural significance and was considered to be the symbol of longevity.

There is a custom of climbing mountains during the Frost's Descent season. Climbing high mountains can make people keep healthy. During the period, North China will be expecting white clouds, red leaves and early frost, which makes people relaxed and happy. The high sky and the red maple leaves are indeed glorious sights when people ascend the mountain and look afar. However, there are some matters about mountain climbing needing attention. First, avoid climbing mountain in the cold morning or

evening. When climbing, people must be calm and walk slowly to prevent sprained waist and leg. Second, don't walk too fast down the hill to avoid knee injury or muscle strain. In the process of climbing, people should adapt to changes in temperature by wearing or taking off clothes. Third, do not sit on the wet ground and places with draughts when you rest. Finally, do not strip off the clothes and hat but slightly loose button when sweat so as not to get the cold. For the elderly, a walking stick should be brought to save energy and promote safety. When climbing a mountain, people should pay attention to the stones under the foot to avoid missing the footing. While walking on a steep slope, people had better take a "Z" shaped route to climb, which can make it gentle slope.

十一月

十一月有立冬、小雪两个季节。

立冬

1. 日期计算

计算公式：(Y×D+C)-L

公式解读：Y= 年数的后 2 位，D=0.2422，L= 闰年数，21 世纪 C=7.438，20 世纪 C=8.218。

举例说明：2088 年立冬日期 =(88×0.2422+7.438)-(88÷4)=

28-22=6,即 11 月 6 日。

例外:2089 年的计算结果加 1 日。

2. 节气特点

立冬是二十四节气中的第十九个节气。一般在每年的 11 月 7 ~ 8 日,此时太阳运行到黄经 225°。我国古时民间习惯以立冬为冬季的开始。《月令七十二候集解》说:“立,建始也。”又说:“冬,终也,万物收藏也。”意思是说,秋季作物全部收晒完毕,收藏入库,动物也已藏起来准备冬眠。因此,立冬不仅仅代表着冬天的来临,还表示万物收藏,规避寒冷的意思。

我国古代将立冬分为三候:“一候水始冰;二候地始冻;三候雉入大水为蜃。”此节气水已经能结成冰;土地也开始冻结;三候“雉入大水为蜃”中的“雉”即指野鸡一类的大鸟,蜃为大蛤,立冬后,雉鸟蛰

伏，而外壳与野鸡的线条及颜色相似的大蛤却在此时大量繁殖，故古人认为野鸡到立冬后便变成大蛤了。我国民间以立冬为冬季之始，事实上此时还不算入冬。气象学上规定，连续5天日平均气温低于10℃以下则视为冬季。但因为我国幅员辽阔，各地气候变化不一。最北部的黑龙江漠河及大兴安岭以北地区，早在9月上旬就已进入冬季，首都北京于10月下旬也已一派冬天的景象，而长江流域的冬季要到小雪节气前后才真正开始。

立冬前后，我国大部分地区降水显著减少，空气一般渐趋干燥，土壤含水较少。雪山上的雪已不再融化。在华北等地往往出现初雪，往往需要特别关注。长江以北和华南地区的雨日和雨量均比江南地区要少。此时，降水的形式出现多样化，有雨、雪、雨夹雪、霰、冰粒等。当有强冷空气影响时，江南也会下雪。

3. 节气习俗

【饮食】在我国南方有立冬吃甘蔗的习俗。甘蔗能成为“补冬”的食物之一，是因为民间素来有“立冬食蔗齿不痛”的说法，意思是立冬之后的甘蔗已经成熟，吃了不上火，这个时候“食蔗”既可以保护牙齿，还可以起到滋补的功效。

羊肉性温热，常吃容易上火。因此，吃羊肉时搭配凉性和甘平性的蔬菜，能起到清凉、解毒、去火的作用。凉性蔬菜一般有冬瓜、丝瓜、油菜、菠菜、白菜、金针菇、蘑菇、莲藕、茭白、笋等；而红薯、土豆、香菇等是甘平性的蔬菜。吃羊肉时最好搭配豆腐，它不仅能补充多种微量元素，还能起到清热泻火、除烦止渴的作用。而羊肉和萝卜做成一道菜，则能充分发挥萝卜性凉，可消积滞、化痰热的作用。

我国北方素来有立冬吃饺子的传统。饺子的原名据称叫“娇

耳”,是医圣张仲景首先发明的。相传他在立冬那天施舍“祛寒娇耳汤”为乡亲医治冻疮。他把羊肉、辣椒和一些驱寒药材放在锅里熬煮,然后将羊肉、药物捞出来切碎,用面包成耳朵样的“娇耳”。煮熟后,分给来求药的人每人两只“娇耳”,一大碗肉汤。人们吃了“娇耳”,喝了“祛寒汤”,浑身暖和,两耳发热,冻伤的耳朵都治好了。后人学着“娇耳”的样子,包成食物,也叫“饺子”或“扁食”。自此民间便有了“立冬不端饺子碗,冻掉耳朵没人管”的说法。

立春至春分、立秋至秋分,要慎喝白酒。这两段时间喝白酒,易引起内脏燥热。

【活动】有些地方庆祝立冬的方式有了创新,在立冬之日,冬泳爱好者们就会用冬泳这种方式迎接冬天的到来。冬泳无论在北方还是南方,都是受人们喜爱的一种锻炼身体的方法。

小雪

1. 日期计算

计算公式:(Y×D+C)-L

公式解读:Y= 年数的后 2 位, D=0.2422, L= 闰年数,21 世纪 C=22.36,20 世纪 C=23.08。

举例说明:2088 年小雪日期 =(88×0.2422+22.36)-(88÷4)=43-22=21,即 11 月 21 日。

例外:1978 年的计算结果加 1 日。

2. 节气特点

小雪是二十四节气中的第二十个节气,时间在 11 月 22 或 23 日。

此时太阳到达黄经 240°。小雪是反映天气现象的节令，代表降雪季节，尤其是北方。《月令七十二候集解》中记载："十月中，雨下而为寒气所薄，故凝而为雪。小者未盛之辞。"小雪阶段比入冬阶段气温低。到了小雪节气，意味着我国华北地区将有降雪。冷空气使我国北方大部地区气温逐步达到 0℃以下。黄河中下游平均初雪期基本与小雪节气一致。虽然开始下雪，一般雪量较小，并且夜冻昼化。我国古代将小雪分为三候："一候虹藏不见；二候天气上升地气下降；三候闭塞而成冬。"由于天空中的阳气上升，地中的阴气下降，导致天地不通，阴阳不交，所以万物失去生机，天地闭塞而转入严寒的冬天。

在立冬节气，我国西北、东北的大部分地区已经开始飘雪，到了小雪节气，意味着华北地区将有降雪。只有在云层内及云下气层的

气温都在0℃以下时，降水形式才会由雨变成雪。如果说立冬节气标志着我国北方大部地区进入冬季的话，到了小雪节气，冷空气的直接表现就是使这些地区的气温逐步降到0℃以下，出现降雪。

3. 节气习俗

【饮食】在我国，腌制腊肉已有几千年的历史。各地都有腌制腊肉的传统习惯，非常普遍。小雪时节，民间有“冬腊风腌，蓄以御冬”的习俗。小雪后气温急剧下降，天气变得干燥，正是加工腊肉的好时候。小雪节气后，一些农家开始动手做香肠、腊肉，等到春节时正好享受美食。过去受条件所限，人们想了很多办法保存食物，腌腊肉便是其中一种。腌了腊肉，寒冬腊月不用出门就能享用到美食。时至今日每到小雪节气的时候，人们也要到市场上挑选肉质上好的猪、鸡、鸭、鱼等带回家腌腊味，品腊味，增添岁暮的温馨气氛。不过，腊肉不能多吃，因为腊肉的脂肪含量非常高，并含有大量的盐，而且含有大量的致癌物质亚硝酸，所以腊肉对很多人，尤其是高血脂、高血糖、高血压等慢性疾病患者和老年朋友而言，不是一种适宜的食物。

南方某些地方还有农历十月吃糍粑的习俗。古时，糍粑是传统的节日祭品，最早是农民用来祭牛神的供品。有俗语“十月朝，糍粑禄禄烧”，就是指的祭祀。

【活动】《诗经·国风》有言：“十月获稻，为此春酒，以介眉寿。”可见酿酒多在冬季，因农事已毕，谷物收获，而岁末祭祀活动频繁，酒的用途比较广泛。

November

There are two solar terms in November—Start of Winter and Slight Snow.

Start of Winter

1. Calculation Formula of Start of Winter

The calculation formula: (Y×D+C)-L

The interpretation of formula: Y= the latter two digits of the year number, D=0.2422, L= the number of leap year, 21st C=7.438, 20th C=8.218("C" stands for century.).

For example, the date of the Start of Winter in 2088=(88×0.2422+7.438)-(88÷4)=28-22=6. That is to say, the day will fall on November 6th.

Exception: The date of the Start of Winter in 2089 will be the day after the calculating result.

2. Features of Start of Winter

Start of Winter (*Lidong*) is the 19th term among the Twenty-four Solar Terms and begins around November 7th and 8th, when the sun is exactly at the celestial longitude of 225°. This solar term is considered the beginning of the winter on Chinese lunar calendar. *A Collective Interpretation of the Seventy-two Pentads* records, "*Li* means the beginning while *dong* stands for winter and an end." It means the winter is coming and crops harvested in autumn should be stored up.

In ancient times, the Start of Winter solar term has been divided into three pentads, "Water begins to freeze, the earth begins to harden and pheasants enter the water for clams." This describes the days after the Start of Winter as the "three periods of waiting"—waiting for the water to turn into ice, waiting for the earth to freeze, and waiting for the pheasants to disappear and big clam of similar colors to show up at the seaside. Just as its name indicates, *Lidong* declares the beginning of winter in Chinese culture, yet not necessarily the meteorological winter. Since China has a vast territory, the temperature difference between the north and south is great. Meteorologists define winter as the period with an average temperature below 10°C for five continuous days. While some northeastern regions in Heilongjiang Province may enter meteorological winter as early as in September, it usually takes much longer for it to reach Beijing, which is not until late October. It reaches the Yangtze River Delta by the Slight Snow.

Before and after the Start of Winter, rainfall begins to decrease significantly in most parts of China. The air becomes drier and the soil contains less water. The snow on the snow-capped mountains is no longer melting. In North China, there are often first snow in this season. It often requires special attention. The areas in north of the Yangtze River and South China have less rainy day and rainfall than those in the south of the Yangtze River. At this time, the forming of rainfall has diversified: rain, snow, sleet, graupel, ice grains, etc. When there is strong cold air, it will snow in the south of the Yangtze River.

3. Customs of Start of Winter

【Diet】There is a custom of eating sugarcane in the south of China. The reason why sugarcane becomes one of the tonics in winter is that eating sugarcane on the Start of Winter doesn't lead to toothache. In this season, sugarcane has matured. People won't suffer from excessive internal heat if they eat sugarcane. At this time, eating sugarcane can both protect teeth and nourish the body.

Mutton is warm in nature. It is very easy to suffer from excessive internal heat if people often eat mutton. Therefore, when people eat mutton, it is necessary to have cold and neutral vegetables, which can play a role in cooling, detoxification and clearing heat. Vegetables of cold nature generally include white gourds, loofahs, rape, spinach, cabbages, needle mushrooms, mushrooms, lotus roots, wild rice stems and bamboo shoots. Sweet potatoes, potatoes and shiitakes belong to vegetables of neutral nature. The tofu is a good side dish when people eat mutton. It can not only supplement a variety of trace elements, but also play a role in clearing heat and quenching thirst. The dish of mutton and carrots mixed together can give full play to the effects of relieving food retention and reducing phlegm due to the cold nature of carrots.

On the day of Start of Winter, there is a tradition to eat dumplings in North China. A legend says that at the end of the Eastern Han Dynasty, "Medical Saint" Zhang Zhongjing saved many people from a typhoid

epidemic and their ears' from being frost bitten around Start of Winter. He cooked mutton, hot peppers and herbs to dispel the cold and increase body heat. He wrapped these ingredients into a dough skin and made them into an ear shape. Since then, people have learned to make the food which became known as "dumpling". Today there is still a saying that goes, "Eat dumplings on Start of Winter, or your ears will be frostbitten."

From the Start of Spring to the Spring Equinox and from the Start of Autumn to the Autumn Equinox, be careful when drinking alcohol. Drinking alcohol during these two periods could lead to heat in internal organs easily.

【Event】In some places, the way of celebrating the Start of Winter has been innovated. On the day of the Start of Winter, winter swimmers use the method of winter swimming to welcome the arrival of winter. Winter swimming is a favorite way to build up body in winter, both in the north and south.

Slight Snow

1. Calculation Formula of Slight Snow

The calculation formula: $(Y \times D+C)-L$

The interpretation of formula: Y=the latter two digits of the year number, D=0.2422, L=the number of leap year, 21st C=22.36, 20th C=23.08("C" stands for century.).

For example, the date of Slight Snow in 2088=(88×0.2422+22.36)-

(88÷4)=43-22=21. That is to say, the day will fall on November 21st.

Exception: The date of Slight Snow in 1978 would be the day after the calculating result.

2. Features of Slight Snow

Slight Snow (*Xiaoxue*) is the 20th term in the Twenty-four Solar Terms. It falls on November 22nd or 23rd every year, when the sun reaches the celestial longitude of 240°. The Slight Snow refers to the time when it starts to snow, mostly in China's northern areas, and the temperature continues to drop. *A Collective Interpretation of the Seventy-two Pentads* explains, "Slight Snow lies in the 10th lunar month. Rain falls and turns into snow when it meets the cold air. The snow is light." The temperature of most areas in the north drops to 0°C and below. It is colder than before. While in the lower and middle reaches of the Yellow River, the average time of the first snow is in the Slight Snow. The snow is light and frozen at night, but melts quickly during the day. In ancient times, the Slight Snow has been divided into three pentads: "Rainbows are concealed from view, the *qi* of the sky ascends, the *qi* of the earth descends, and closure and stasis create winter." It was believed that rainbows were the results of yin and yang energy mixing; winter, being dominated by yin, would not present rainbows. The end of mixing between sky and earth, yin and yang, leads to the dormancy of winter.

On the Start of Winter, most parts of the northwest and northeast of China have begun to snow. The arrival of Slight Snow means that there

will be snowfall in North China. It is well known that only when the temperature in the cloud and under the cloud is below 0°C, the form of precipitation changes from rain to snow. If the Start of Winter marks the entry of winter in most parts of northern China, the direct manifestation of cold air in Slight Snow is to gradually reduce the temperature in these areas to below 0°C, and snow falls.

3. Customs of Slight Snow

【Diet】In our country, making cured meat has a history of thousands of years. After the Slight Snow, the temperature declines sharply and the air becomes dry. It is the best time to start making cured meat. Until the Chinese Spring Festival, it will be well made and enjoyed. In the past, when storage conditions were poor, people created many ways to store food. Cured meat is one of them. Thus even in the bitter winter, the whole family could enjoy meat without going out. Today, when the Slight Snow arrives, people also go to the market to select the best pork, chicken, duck, fish, etc. They pickle these ingredients and enjoy the delicacy, which adds the warm atmosphere of the year. However, the cured meat cannot be eaten too much, because it is very high in fat content and contains a large amount of salt. It also contains a large amount of carcinogenic nitrite, so cured meat is not good for many people, especially for the elderly and patients with chronic diseases such as hyperlipidemia, hyperglycemia and hypertension.

In some regions of South China, there is also the custom of eating glutinous rice cakes in the tenth month of the lunar calendar, which

is around the Slight Snow. In ancient times, glutinous rice cake was a traditional festival sacrifice offering to the God of Cattle by peasants. There is a saying that "in the tenth month, burn the cooked glutinous rice pounded into paste", which refers to the sacrifice.

【Event】*Guofeng, The Book of Songs* says, "In the tenth month, get rice and make spring wine for longevity." It can be seen that wine making is done mostly in the winter. Because the agricultural work has been completed and the grain has been harvested, the ritual activities at the end of the year are frequently held and the use of wine is extensive.

十二月

十二月有大雪、冬至两个节气。

大雪

1. 日期计算

计算公式：(Y×D+C)-L

公式解读：Y= 年数的后 2 位，D=0.2422，L= 闰年数，21 世纪 C=7.18，20 世纪 C=7.9。

举例说明：2088 年大雪日期 = (88×0.2422+7.18)-(88÷4)=28-22=6，即 12 月 6 日。

例外：1954 年的计算结果加 1 日。

2. 节气特点

大雪是二十四节气中的第二十一个节气，通常在每年12月7日或8日，此时太阳黄经达255°。“大雪”并非是说雪很大，而是指大雪时节后，降雪概率相比小雪更大，降雪范围更广。据《月令七十二候集解》：“大雪，十一月节。大者，盛也。至此而雪盛矣。”我国古代将大雪分为三候：“一候鹖鴠不鸣；二候虎始交；三候荔挺出。”这是说此时因天气寒冷，寒号鸟也不再鸣叫了；由于此时是阴气最盛时期，正所谓盛极而衰，阳气已有所萌动，所以老虎开始有求偶行为；“荔挺”为草的一种，也感到阳气的萌动而抽出新芽。

大雪时节，除华南和云南南部无冬区外，我国辽阔的大地已披上冬日盛装。东北、西北地区平均气温达-10℃以下，黄河流域和华北地区气温也稳定在0℃以下。此时，北方地区的降雪可能会持续一整天，会压断树枝、阻塞道路，俨然一片“千里冰封，万里雪飘”的

北国风光。在南方，有时也会雪花飞舞，世界变成白茫茫的一片。人们常说，“瑞雪兆丰年”。严冬积雪覆盖大地，庄稼如同盖上了厚厚的棉被，过冬的害虫会被低温冻死，给冬作物创造了良好的越冬环境。大雪时节，北方田间管理已很少，冬小麦停止了生长。江淮及以南地区小麦、油菜仍在缓慢生长，要注意施肥，为安全越冬和来春生长打好基础。

3. 节气习俗

【饮食】大雪节气是进补的最佳时节，可以适当多吃些牛肉、兔肉和羊肉等，像面红上火、干咳、皮肤干燥等阴虚之人应以防燥、滋肾润肺为主，可食用柔软甘润的食物，如牛奶、豆浆、鸡蛋、鱼肉等，忌食燥热食物，如辣椒、胡椒、茴香等。如果是经常面色苍白、四肢乏力、易疲劳怕冷等阳虚之人，应食用温热、熟软的食物，如豆类、大枣、南瓜、韭菜、芹菜、鸡肉等，忌食黏、干、硬、生冷的食物。

【活动】滑冰是冬季游戏之一，古时称为“冰戏”。北方严寒，河流冻得坚实，滑冰最为流行。男女经常穿着冰鞋，动作轻捷如飞。善于溜冰的多不用拄杖，技巧高超的更能做出种种花样。清代乾隆帝和慈禧太后冬月经常在北海漪澜堂观赏冰戏。南方气候温暖，冰冻得不够结实，因此滑冰的很少。

冬至

1. 日期计算

计算公式：(Y×D+C)-L

公式解读：Y= 年数的后 2 位，D=0.2422，L= 闰年数，21 世纪

C=21.94,20 世纪 C=22.60。

举例说明：2088 年冬至日期 =(88×0.2422+21.94)-(88÷4)=43-22=21,即 12 月 21 日。

例外：1918 年和 2021 年的计算结果减 1 日。

2. 节气特点

冬至是二十四节气中的第二十二个节气,时间在每年的公历 12 月 21 ~ 23 日。冬至这天,太阳运行至黄经 270° ,太阳直射地面的位置到达一年的最南端,太阳几乎直射南回归线。因此,冬至日是北半球各地一年中白昼最短的一天,也是全年正午太阳高度最低的一天。《月令七十二候集解》有言:“十一月十五日,终藏之气,至此而极也。”古人对冬至的说法是:阴极之至,阳气始生,日南至,日短之至,日影长之至,故曰“冬至”。所以古时有“冬至一阳生”的说法,意思

是说从冬至开始，阳气慢慢地回升，白昼一天比一天长，所以民间又有“吃了冬至面，一天长一线”之说。

我国古代将冬至分为三候：“一候蚯蚓结；二候麋角解；三候水泉动。”传说蚯蚓是阴曲阳伸的生物，此时阳气虽已生长，但阴气仍然十分强盛，土中的蚯蚓仍然蜷缩着身体；麋与鹿同科，却阴阳不同，古人认为麋的角朝后生，所以为阴，而冬至一阳生，麋感阴气渐退而解角；由于阳气初生，所以此时山中的泉水可以流动并且温热。冬至前后，虽然北半球日照时间最短，接收的太阳辐射量最少，但这时地面积蓄的热量还可提供一定的补充，故这时气温还不是最低。冬至过后，气温在一段时间内会继续下降，进入一年中最冷的时期。我国除少数海岛和海滨局部地区外，1月都是最冷的月份，故民间有“冬至不过不冷”之说。

从气候上看，冬至预示着寒冬的来临。此时，西北地区平均气温普遍在0℃以下，南方地区也只有6℃至8℃左右。另外，冬至开始“数九”，冬至日也就成了“数九”的第一天。不过，西南低海拔河谷地区，即使在当地最冷的1月上旬，平均气温仍然在10℃以上，真可谓秋去春平，全年无冬。

3. 节气习俗

【饮食】由于地方差异性，不同地方冬至习俗不一样，一般北方地区冬至是吃饺子的，南方人在冬至这一天习惯吃汤圆、馄饨。关于北方吃饺子，还有一个小段子，传说不吃饺子会冻掉耳朵的；而在南方，冬至一定要吃汤圆，汤圆预示着团团圆圆，还有“吃了汤圆大一岁”的说法。

冬季是便秘的多发季节，早晨吃猕猴桃可治便秘。中医认为，

这个时节干燥的气候容易在人体内诱发燥邪,耗伤人体津液,便秘就是这一现象的主要体现。临床上总是建议便秘的人采取食疗的方法,不仅效果好,而且没有副作用。多吃水果就是食疗中最有效的一种。

入冬后,很多人常感到皮肤干燥、头晕嗜睡,反应能力降低,这时如果能吃些生津止渴、润喉去燥的水果,会使人顿觉清爽舒适。甘蔗是含水分很高的水果,它的水分含量为84%,在干燥的冬天补充水分是必不可少的,其次甘蔗的含铁量在众多水果中可是名列前茅的。它作为清凉的补剂,对于治疗低血糖、大便干结、小便不利、口渴、反胃呕吐以及肺燥引发的咳嗽气喘等病症有一定的疗效。不过,由于甘蔗性寒,脾胃虚寒的人不宜食用。

【活动】冬至作为节日已有2 500年的历史。早在商周时期,明确规定冬至前一天为岁终之日,冬至节相当于春节。《周礼·春官·神仕》上记载:"以冬日至,致天神人鬼。"专门过"冬至节"始于汉代,盛于唐宋,相沿至今。汉朝最开始将冬至日定为节日"冬节",官民放假举行"贺冬"庆典。家家户户都像过年一样的团聚,祭奠祖先与神明,吃上一顿丰盛的团圆饭。唐宋时期,冬至是祭天祀祖的日子,皇帝在这天要到郊外举行祭天大典,百姓在这一天要向父母尊长祭拜。一直到清朝,《清嘉录》明确说明"冬至大如年",冬至节的意义可见一斑。

冬至日祭祀祖先是全国各地普遍的习俗,又称"冬祭"。同姓同宗者于冬至前后上坟,给祖坟添土或竖碑。室内祭奠,则在家祠举行。族人集聚家祠,照长幼之序祭拜祖先,俗称"祭祖"。祭典之后,还会大摆宴席,招待前来祭祖的宗亲们。

December

There are two solar terms in December—Great Snow and Winter Solstice.

Great Snow

1. Calculation Formula of Great Snow

The calculation formula: (Y×D+C)-L

The interpretation of formula: Y=the latter two digits of the year number, D=0.2422, L=the number of leap year, 21st C=7.18, 20th C=7.9("C" stands for century.).

For example, the date of Great Snow in 2088=(88×0.2422+7.18)-(88÷4)=28-22=6. That is to say, the day will fall on December 6th.

Exception: The date of Great Snow in 1954 would be the day after the calculating result.

2. Features of Great Snow

Great Snow (*Daxue*) is the 21st term in the Twenty-four Solar Terms and falls on December 7th or 8th every year, when the sun reaches the celestial longitude of 255°. The Great Snow actually means it is getting colder, and there is frequent snowfall compared with that of Slight Snow; and it does not mean that there will be a heavy snowfall. *A Collective Interpretation of the Seventy-two Pentads* mentions: "Great Snow lies in the 11th lunar month. 'Great' means a large scale." In ancient times, there is

three pentads of the Great Snow: " In the first pentad, the jie-bird ceases to crow; the second, tigers begin to mate; and the third, buds appear on *liting*." The *jie* is a bird, similar to the pheasant, which is believed to be aggressive and combatant. As winter progresses, even this active bird slows and ceases to crow. Though the yin is still very powerful, the yang begins and tigers begin to mate. Buds appear on *liting*, a kind of grass.

During the Great Snow season, most parts of China, except for South China and the southern part of Yunnan Province, are usually covered with snow. In the northeastern and northwestern regions, the average temperature has dropped below -10℃. In the Yellow River region and North China, the temperature is stable at below 0℃ . The snow in North China may last a whole day, breaking tree branches and blocking the road. The natural scenery is "hundreds miles locked in ice, thousands miles of whirling snow". In the south, the snowflakes fly, and the world turns white. About the snow, a proverb goes, "A timely snow promises a good harvest." If the snow is very heavy, it means the next year will be a harvesting year. As the snow covers the ground, the snow melts and soaks into the soil and pests living through the winter will be killed by the low temperature. There are little field management at this time. Winter wheat has stopped growing. In the Yangtze-Huaihe Region and its southern areas, wheat and cole still grow, but in a very slow pace. This period requires fertilization, which can lay the good foundation for their safely wintering and growth in the coming spring.

3. Customs of Great Snow

【Diet】Great Snow is the best season for nourishing the body.

People can eat more beef, rabbit meat and mutton, which are excellent for promoting blood circulation and providing protection against the cold. Those who are deficient in yin suffering from excessive internal heat, dry coughing and dry skin should clear dryness, nourish the kidneys and moisten the lungs. They can eat soft and moisturizing food, such as milk, soybean milk, eggs, fish and so on. Do not eat hot food like chili, pepper, fennel, etc. If someone who is deficient in yang often has such symptoms as pale complexion, weak limbs, fatigue and sensation of chill, he should eat warm and soft food like beans, jujube, pumpkin, leeks, celery, chicken and so on. Do not eat sticky, dry, hard and cold food.

【Event】Skating is one of the winter games. It was called "ice opera" in ancient times. In the cold north, the river is frozen solid and skating is the most popular activity. Men and women often wear skates and look like a bird when skating. Those who are good at skating don't need a cane and the skating aces can even make a variety of tricks. In the Qing Dynasty, Emperor Qianlong and Empress Dowager Cixi often watched ice operas in Yilan Temple of Beihai. Since the southern climate is not cold enough to freeze the river, there are very little people skating.

Winter Solstice

1. Calculation Formula of Winter Solstice

The calculation formula: (Y×D+C)-L

The interpretation of formula: Y=the latter two digits of the year

number, D=0.2422, L=the number of leap year, 21st C=21.94, 20th C=22.60("C" stands for century.).

For example, the date of the Winter Solstice in 2088=(88×0.2422+21.94)-(88÷4)=43-22=21. That is to say, the day will fall on December 21st.

Exception: The date of the Winter Solstice in 1918 would be the day before the calculating date.

2. Features of Winter Solstice

Winter Solstice (*Dongzhi*) is the 22nd solar term and falls around December 21st to 23rd each year. On the day, the sun reaches 270° of longitude, and shines directly on the Tropic of Capricorn. On the day of Winter Solstice, the Northern Hemisphere experiences the shortest period of daylight and the longest night of the year, when the sun is at its lowest daily maximum elevation in the sky. From then on, the days become longer and the nights become shorter. In *A Collective Interpretation of the Seventy-two Pentads*, it says, "On the 15th day of the 11th month in lunar calendar, the hidden *qi* reaches its peak." The Winter Solstice comes on the fifteenth day after the Great Snow. It is in the middle of the eleventh month. The yin goes comes to an end and the yang begins. The sun goes to the south. The days begin to draw out after the Winter Solstice day.

In ancient times, there are three pentads of the Winter Solstice: "In the first pentad, earthworms curl." It is said the earthworms curl in the yin and extend in the yang. Though the yang has begun, the yin is still very

powerful. Thus the earthworm is still curling. “In the second, the horns of *mi* (Pere David’s Deer) disappear.” Pere David’s Deer belongs to yin, while deer belongs to yang. The ancient Chinese believed that because the horns of *mi* grow backward, it belongs to the yin. During the Winter Solstice, the yang begins, so the horns gradually disappear. “In the third pentad, the spring flows.” As the yang begins, the water of springs in mountains begins to flow and become warm. Around the day of the Winter Solstice, the duration of sunshine is shortest and the heat dissipated from the ground is more than the heat it receives. However, it is not the coldest time because of the reserved heat in the soil. After the Winter Solstice, the temperature decreases gradually and the coldest season comes. Except the islands and areas along the sea, January is the coldest month in China. Thus, there is a saying that “if you are not in the Winter Solstice season, you will not feel cold”.

Astronomically, the Winter Solstice marks the coming of the coldest season in the year. During this period, the average temperature in the northwest areas of China is below 0℃, and it is about 6℃ to 8℃. The Chinese make nine days into a clump, and around the third nine-day, it is the coldest. In the low-altitude valley of the southwest, however, the average temperature is above 10℃ even in the coldest month. In that area, the spring comes after autumn and there is no winter at all.

3. Customs of Winter Solstice

【Diet】Traditions and customs for the Winter Solstice vary in

different areas of China. In the north of China, eating dumplings is an essential thing to do. By eating dumplings, people think they can avoid their ears from being frostbitten as the dumplings look like people's ears. And in South China, people usually eat *tangyuan* and wonton to celebrate the day. *Tangyuan*, a kind of stuffed small dumpling ball made of glutinous rice flour, symbolizes reunion and unity. People think after eating it, they will be one year older.

It is easy to suffer from constipation in winter. Eating kiwi in the morning can cure constipation. Traditional Chinese Medicine believes that the dry climate at this time of the year can induce pathogenic dryness in the human body easily. It is apt to consume the body's fluid and constipation is the main manifestation of this phenomenon. From a clinical perspective, it is always recommended that people with constipation should undergo dietary therapy, which not only has good result, but also has no side effects. Eating more fruits is one of the most effective dietary therapies.

In winter, many people often feel dry, dizzy and sleepy. Their response ability is reduced. If they can eat some fruits that can help produce body fluids and slake the thirst, moisten throat and clear dryness, people will feel refreshed and comfortable. Sugarcane is a kind of fruit with high moisture content of 84 percent. It is indispensable to supplement water in dry winter. Furthermore, the iron content of sugarcane is in the lead among many fruits. As a refreshing supplement, it has certain effect for the treatment of hypoglycemia, dry stool, dysuria, thirst, nausea and vomiting,

cough and asthma caused by lung dryness. However, due to the cold nature of sugarcane, people with weak spleen and stomach should not eat it.

【Event】Winter Solstice Festival has a far-reaching history of about 2 500 years. As early as the Shang and Zhou Dynasty, the day before Winter Solstice was set as the last day of the year, and Winter Solstice was the "Spring Festival". People worshipped the gods on the first day of the Winter Solstice, according to *The Rites of Zhou*. The Winter Solstice became a festival during the Han Dynasty and thrived in the Tang and Song dynasties. People in the old times attached great importance to the festival. The Han people regarded Winter Solstice as a "Winter Festival", so the celebratory activities were officially organized. On this day, both officials and common people would have a rest. Relatives and friends presented to each other delicious food. In the Tang and Song dynasties, the Winter Solstice was a day to offer sacrifice to Heaven and ancestors. Emperors would go to suburbs to worship the Heaven, while common people offered sacrifice to their deceased parents or other relatives. The Qing Dynasty even had the record that "Winter Solstice is as formal as the Spring Festival", showing the great importance attached to this day.

It is a popular tradition to worship and commemorate ancestors on the day of Winter Solstice. Around the day, people of the same surname or family clan visit, clean and make offerings at ancestral graves, or gather at their ancestral temples to worship their ancestors in age order. After the sacrificial ceremony, there is always a grand banquet.